ENVIRONMENTAL STATUS OF INDIA

ENVIRONMENTAL STATUS OF INDIA

Editors
Sukumar Devotta
C.V. Chalapati Rao

ATLANTIC
PUBLISHERS & DISTRIBUTORS (P) LTD

Published by

ATLANTIC

PUBLISHERS & DISTRIBUTORS (P) LTD

B-2, Vishal Enclave, Opp. Rajouri Garden,
New Delhi-110027
Phones : +91-11-25413460, 25429987, 25466842

Sales Office
7/22, Ansari Road, Darya Ganj,
New Delhi-110002
Phones : +91-11-23273880, 23275880, 23280451
Fax : +91-11-23285873
web : www.atlanticbooks.com
e-mail : info@atlanticbooks.com

Branch Office
5, Nallathambi Street, Wallajah Road,
Chennai-600002
Phones : +91-44-64611085, 32413319
e-mail : chennai@atlanticbooks.com

ISBN 978-81-269-0698-7

Printed in India
at Nice Printing Press, Delhi

Preface

Environmental pollution has assumed alarming proportions. Its adverse impact on the everyday lives of people is increasing in magnitude and intensity. That more and more people are becoming aware of this hazard is evident from the fact that National Environmental Engineering Research Institute is inundated with queries relating to various aspects of environmental pollution. This book has been prepared to present the environmental status of India, study various pertinent issues and suggest measures to mitigate the harmful effects of environmental pollution.

The book is divided into seven main sections, each dealing with an important aspect of environmental pollution. The Air Pollution section describes the different sources and types of air pollution that we face, the status of air quality, the concerns and inadequacy in management and control and their related health impacts. It also gives an insight into the air quality standards in India and suggests how we could tackle the future challenges.

It is rightly said that the reason we have life on Earth is that we have water. Unfortunately, however, for many of us, water is simply what flows from our taps; we think little beyond this. We have lost a sense of respect for our water sources like rivers, streams, lakes, ponds, tanks, wells, etc. The complex working of wetlands and intricate web of life that water supports are continuously being disturbed by unsustainable human practices. The section on Water takes up all these issues and studies water resources, their distribution across India and the major contaminants that pollute water. Millions of people across the country have to consume water in which the levels of arsenic or fluoride are far above acceptable levels, thus exposing themselves to serious health hazards. Solving the problem of groundwater contamination through bioremediation and meeting future challenges like decrease in the level of underground water have also been discussed in this section.

One of the ill effects of economic growth and rapid industrialization is the generation of wastewater, heavily contaminated with chemicals. One section of the book exclusively deals with the problem of wastewater generated by industries, municipal sewage, and its treatment through microbial and physico-chemical options.

The section on Solid Waste and Hazardous Waste categorises the types of waste generated and suggests different treatment options thereof. The challenges of disposal of hazardous wastes have also been discussed.

Pollution related problems in soil and land use have been analysed in Land Environment section while the effects of changing biodiversity of an ecosystem have been taken up in the Biodiversity section.

This book is not meant to be an encyclopaedia on environmental issues, nor a textbook for students. It is a compilation of important aspects that are required for quick reference on the subject. Experts from NEERI, with vast experience in these areas, discuss the issues being faced by us today and suggest solutions for the future challenges. We hope that readers will find this a useful anthology.

We are thankful to all the authors for their contribution to this book. We would like to thank Mrs. Anuradha Dorlikar for her diligent typing and patience with many revisions of the draft.

SUKUMAR DEVOTTA
C.V. CHALAPATI RAO

Contents

Contributors

C.V. Chalapati Rao, Scientist, Air Pollution Control Division, NEERI, Nagpur.
Ph: +91-712-2249895, E-mail: *cvc_rao@neeri.res.in*

S.K. Goyal, Scientist, Air Pollution Control Division, NEERI, Nagpur.
Ph: +91-712-2249886, E-mail: *sk_goyal@neeri.res.in*

Rakesh Kumar, Scientist & Head, NEERI Zonal Laboratory, Mumbai.
Ph:+91-022-24974607, E-mail: *rakeshme@rediffmail.com*

S.M. Tamhane, Scientist, Air Pollution Control Division, NEERI, Nagpur.
Ph: +91-712-2249886, E-mail: *sm_tamhane@neeri.res.in*

Padma S. Rao, Scientist, Air Pollution Control Division, NEERI, Nagpur.
Ph: +91-712-2249886, E-mail: *ps_rao@neeri.res.in*

N.P. Thacker, Scientist & Head, Analytical Instrumentation Division, NEERI, Nagpur.
Ph: +91-712-2249886, E-mail: *np_thacker@neeri.res.in*

Anjali Srivastava, Scientist & Head, NEERI Zonal Laboratory, Kolkata.
Ph: +91-033-24421988, E-mail: *srivastv@vsnl.com*

J.S. Pandey, Scientist, Environmental Risk & Assessment Division, NEERI, Nagpur.
Ph: +91-712-2249886, E-mail: *js_pandey@neeri.res.in*

M.V. Nanoti, Ex-Scientist & Head, Geo Environmental Management Division, NEERI, Nagpur.

Apurba Gupta, Scientist & Head, Environmental Systems Design & Modelling Division, NEERI, Nagpur.
Ph: +91-712-2249990, E-mail: *a_gupta@neeri.res.in*

V.A. Joshi, Scientist, Geo Environmental Management Division, NEERI, Nagpur.
Ph: +91-712-2249886, E-mail: *va_joshi@neeri.res.in*

M.P. Patil, Scientist, Environmental Biotechnology Division, NEERI, Nagpur.
Ph: +91-712-2249886, E-mail: *mp_patil@neeri.res.in*

Sadhna Rayalu, Scientist & Head, Environmental Materials Unit, NEERI, Nagpur.
Ph: +91-712-2247828, E-mail: *s_rayalu@neeri.res.in*

T. Nandy, Scientist & Head, Waste Water Technology Division, NEERI, Nagpur.
Ph: +91-712-2249763, E-mail: *t_nandy@neeri.res.in*

N.N. Rao, Scientist, Waste Water Technology Division, NEERI, Nagpur.
Ph: +91-712-2249886, E-mail: *nn_rao@neeri.res.in*

H.J. Purohit, Scientist & Head, Environmental Geonomics Unit, NEERI, Nagpur.
Ph: +91-712-2249883, E-mail: *hj_purohit@neeri.res.in*

Atya Kaple, Scientist, Environmental Geonomics Unit, NEERI, Nagpur.
Ph: +91-712-2249883, E-mail: *a_kapley@neeri.res.in*

A.A. Juwarkar, Scientist, Environmental Biotechnology Division, NEERI, Nagpur.
Ph: +91-712-2249886, E-mail: *aa_juwarkar@neeri.res.in*

V.P. Deshpande, Ex-Scientist, Solid Waste Management Division, NEERI, Nagpur.

J.K. Bhattacharya, Scientist, Solid Waste Management Division, NEERI, Nagpur.
Ph: +91-712-2249886, E-mail: *jk_bhattacharya@neeri.res.in*

T. Chakrabarti, Scientist & Head, Environmental Biotechnology Division, NEERI, Nagpur.
Ph: +91-712-2249759, E-mail: *t_chakrabarti@neeri.res.in*

A.M. Deshkar, Ex-Scientist, Environmental Biotechnology Division, NEERI, Nagpur.

Rekha Thakre, Ex-Scientist, Geo Environmental Management Division, NEERI, Nagpur.

T.K. Ghosh, Scientist, Environmental Biotechnology Division, NEERI, Nagpur.
Ph: +91-712-2249886, E-mail: *tk_ghosh@neeri.res.in*

AIR POLLUTION

Air Quality Standards and Monitoring in India

1

C.V. Chalapati Rao

Preamble

Air pollution has become a growing problem particularly in urban areas. India has many metropolitan cities, large number of urban areas and vast rural region. There are several sources of air pollution such as power plants, industrial stacks, vehicular traffic, domestic fuel burning, fossil fuel burning and many other anthropogenic activities in urban areas in India. In the rural sector, the major sources of air pollution are domestic fuel burning, chullahs and agricultural emissions such as pollen, biomass burning, etc. As a result, several air pollutants such as suspended particulate matter (SPM), respirable suspended particulate matter (RSPM/PM10), sulphur dioxide (SO_2), oxides of nitrogen (NOx), carbon monoxide (CO), hydrocarbons (HC), volatile organic compounds (VOC), methane (CH_4), benzene, etc.; several specific pollutants such as ammonia (NH_3), chlorine (Cl_2) are released into the air environment. Secondary pollutants such as ozone (O_3) are generated as a result of photochemical reactions in the lower atmosphere.

In order to understand and analyze the problem of air pollution and to formulate an appropriate and effective air quality management strategy for improvement in air quality, a reliable air quality database is a prerequisite and the same calls for stipulation/promulgation of air quality standards and monitoring to check compliance and analyse air quality trends.

Air Quality Standards

Central Pollution Control Board (CPCB), the apex regulatory agency in India, has promulgated National Ambient Air Quality

Standards (NAAQS) along with the method of measurement of different pollutants (CPCB, 1994) since one of the functions and jurisdictions of the Central Pollution Control Board is to lay down standards for quality of air [Section 16 (2) (h) of Air (Prevention and Control of Pollution) Act, 1981]. In the absence of adequate studies/data carried out in India regarding the effect of air pollution on human health, the United States Environmental Protection Agency (USEPA) health criteria were used and the limiting concentration values of various pollutants in the ambient air were evolved considering adequate margin of safety. The rationale and important features of the present set of standards are:

- While evolving the standards, air quality levels with an adequate margin of safety are considered to protect public health, vegetation and monuments, etc.
- The long-term standards (annual average) are designed to protect the population from regular exposure to high levels of pollution. In addition, short-term standards (1 / 8 / 24 hours average) are also prescribed to control acute effects, which result when very high levels of pollution persist for short periods.
- Annual and 24-hourly average standards are fixed for SO_2, NO_2, SPM, RSPM, Lead (Pb) and Ammonia (NH_3). 1 and 8-hourly averages are fixed for CO.
- Different sets of standards are prescribed for (i) industrial areas; (ii) residential, rural and other areas; and (iii) sensitive areas, which are based on the land-use pattern.
- For annual average, annual arithmetic mean of minimum 104 measurements in a year taken twice a week 24-hourly, at uniform intervals, is to be considered.
- 24/8-hourly values should be met 98% of the time in a year and it should not exceed on any two consecutive days.

WHO guidelines are independent of land-use category and CPCB standards are comparable under one or the other category, except for NO_2, which is considered to be a more stringent standard for Indian conditions. National Ambient Air Quality Standards and WHO Guidelines are presented in Table 1 for comparison purpose.

Table 1. National Ambient Air Quality Standards and WHO Guidelines

Pollutants	Average Time	Concentration (μg / m³)			WHO Guidelines
		Industrial	Residential	Sensitive	
SPM	Annual	360	140	70	60-90
	24 hrs	500	200	100	100-150
PM10	Annual	120	60	50	40
	24 hrs	150	100	75	70
SO_2	Annual	80	60	15	40-60
	24 hrs	120	80	30	100-150
NO_2	Annual	80	60	15	150
	24 hrs	120	80	30	
Lead (Pb)	Annual	1.0	0.75	0.50	0.5-1.0
	24 hrs	1.5	1.00	0.75	
Ammonia (NH_3)	Annual	–	100	–	
	24 hrs		400		
CO	8 hrs	5000	2000	1000	10000
	1 hr	10000	4000	2000	30000

Ambient Air Quality Monitoring in India

Ambient Air Quality Monitoring Programme was initiated by CPCB during 1984-85 at the national level to identify the areas in need of restoration of air quality, identification of the nature and extent of pollution control, and to regularly assess the effectiveness of pollution control programmes. This nation-wide programme (known as NAAQM) comprises of 295 monitoring stations covering more than 90 cities/towns in 24 States and 5 Union Territories. Atleast three monitoring stations are operated (industrial, residential, commercial and sensitive area categories) in a city, however, the cities with larger population have more stations. Nearly hundred agencies are involved in air quality monitoring, which include CPCB, State Pollution Control Boards (SPCBs), NEERI, Pollution Control Committees, Universities and Research Institutes throughout the country. The air pollutants monitored regularly at all the sites are SPM, SO_2 and NO_2. SPM is monitored 8-hourly for 24 hours, whereas SO_2 and NO_2 are monitored 4-hourly for 24 hours. Monitoring frequency is twice a week, thus in a year, 104 days observations are available. In addition to criteria pollutants, certain parameters of health significance like PM10, PAHs, trace metals and other gaseous pollutants (NH_3 & H_2S) are also being monitored by NEERI. CO, ozone and benzene are additionally monitored in Delhi and some other cities, but not in all the cities. CPCB analyses, compiles and consolidates all the above data and disseminates in the form of reports and/or through CPCB website. However, quality of data was observed to be one of the constraints, and a need for strong framework for monitoring and analysis of observations has been emphasised.

Air Quality Status—Indian Scenario

Goyal *et al.* (2005) analyzed the air quality status (during 1995-2003) in the residential areas of nineteen major cities having population more than one million (as per 2001 census). In all the cities, the annual mean levels of SPM were in general found to be higher than the CPCB standard of 140 $\mu g/m^3$ during all the years, except at Chennai and Kochi. Lucknow, Kanpur, Delhi and Varanasi were found to be the most polluted cities and the standard exceeded by about 3 times. WHO guideline of 60-90 $\mu g/m^3$ for SPM was found to exceed by 5-6 times in all the cities. High levels of SPM were attributed mainly to the natural sources and to some extent, the anthropogenic sources. However, in general, a decreasing trend was observed from 1995 to 2000, and thereafter a steady increase in the SPM levels for most of the cities. The decline in SPM levels was attributed to enforcement of various measures for air quality improvement and the growing awareness of industries for use of emission control devices. Due to significant variations observed in measured SPM levels (caused by meteorological/atmospheric conditions), it was difficult to use the SPM data as an indicator of air pollution (GOI, 2002).

In all the cities, the concentration levels of the most important parameter of health significance, PM10 were found to exceed the CPCB standard of 60 $\mu g/m^3$. Ahmedabad and Kanpur were found to be the most polluted cities. However, a decreasing trend was observed in Ahmedabad, Bhopal, Kanpur, Kochi, Nagpur and Surat, whereas increasing trends were observed in Bangalore, Delhi, Jaipur and Pune. Various unorganized/unidentified local area source activities may be responsible for such variations.

Annual mean levels of SO_2 were found to be always much below the prescribed standard of 60 μg/m³ during all the years (1995-2003) in all the cities. The WHO guideline of 40-60 μg/m³ for SO_2 was also met. In general, a decreasing trend was observed in all the cities, which was attributed to various pollution control measures, particularly adopted in different industrial sectors, and also due to reduction in sulfur content in the diesel used in automobiles.

Annual mean levels of NO_2 were also found to be mostly below the prescribed standard of 60 μg/m³ during all the years in all the cities, except at Pune (beyond 1998) and Kolkata beyond 2001. A sudden increase from 30 to 66 μg/m³ was observed in Kolkata in 2001 as against in 2000. The NO_2 levels, in general, were found to be between 15 and 40 μg/m³ during 1995-2003, in most of the cities. Kolkata, Delhi and Bangalore had shown an increasing trend after 2001, mainly due to considerable increase in vehicular traffic. However, no specific trend was observed for any city over the years.

Extent of pollution in a city is presented in relative terms of pollution levels (low, moderate, high or critical), which is defined on the basis of exceedence factor (EF). Exceedence factor is the ratio of measured annual mean concentration of a pollutant to the annual standard for that pollutant. The four air quality categories are: (i) EF > 1.5 – critical pollution; (ii) EF between 1 and 1.5 – high pollution; (iii) EF between 0.5 and 1.0 – moderate pollution; and (iv) EF < 0.5 – low pollution. Based on the air quality data and exceedence factor, Delhi, Mumbai, Kolkata, Chennai, Kanpur, Ahmedabad, Bangalore and Pune were identified as highly polluted cities. These cities require more attention for air quality improvement, particularly with respect to particulate matter.

Most Serious Concerns/Emerging Issues

An Expert Committee of Government of India reviewed and analysed the database for different cities and identified several limitations and data gaps. It concluded that the air quality data generated in the network may have certain limitations regarding the probability of variation, biases due to the involvement of large number of monitoring agencies, personnel, equipment for sampling, analysis and data reporting. Because of these limitations, the published air quality data statistics are rather more indicative than absolute (GOI, 2002).

The air quality related studies carried out in India during 1990s were reviewed by TERI to assess the contributions of different sources to urban air pollution, and to pool together the available existing knowledge. It was observed that the data are not comprehensively and statistically analysed for temporal and spatial trends. Many stations no longer represent the site they are originally selected because of the changes in land use pattern, habitation and commercial activities. There is a need for review and possible relocation, if required (TERI, 2001).

A study conducted by World Bank emphasised the need for quality assurance and quality control in air quality monitoring. The areas that call for attention include not only the accuracy and reproducibility of the measurements but also data analysis; timely publication of, and access to, raw data; re-examination of site selection in the light of changes in land use patterns; the actual monitoring frequency; and the positioning of the instruments at a given site (World Bank, 2002). Other limitations include lack of compilation and utilization of data monitored by industries, lack of proper emission inventory, database for polluting sources, emission factors for Indian conditions, lack of proper source apportionment study, chemical profiles, and health impact study. Under such scenario, it is difficult to correlate air quality of a city with various activities and also effectiveness of control measures.

The very criteria of setting uniform land use based standards need to be reviewed due to the growing mixed land use and ever expanding activities. Even if the review suggests continuing with the present system, the area class needs to be defined. The pollutants, relevant in Indian context, may be identified on the basis of prevailing activities (industries, road traffic, fuel usage, etc.) and limiting values may be arrived at considering the dose and duration of exposure of pollutant and associated likely impacts on various types of receptors. Keeping in view all the above, National Ambient Air Quality Standards are being reviewed by Indian Institute of Technology, Kanpur.

Future Challenges/R&D Needs

- Development of assimilative capacity based air quality standards and extensive epidemiological investigations to facilitate development of health related air quality standards.
- Measurement of additional parameters of health significance, viz. PM10 and PM2.5; CO, O_3, PAHs and toxic metals in more cities, also the parameters (viz. sulphates, nitrates, etc.) having damage potential to vegetation, materials, etc. should be included. There is a felt need to strengthen the air quality database in terms of increased monitoring frequency.
- Monitoring studies need to be supplemented by studies of other contributing factors like meteorology, topography, emission inventory, etc., and also there is a need for detailed source apportionment and epidemiological studies linked with air pollution at different socio-economic levels, regions in the country.
- The studies on O_3 made for Delhi, Ahmedabad, Thumba, Pune, etc. show that O_3 levels are observed to be high during morning hours and in winter season. The advent of automobile exhaust in major Indian cities releasing NOx and VOCs are responsible for the generation of O_3 in lower atmosphere. The increase in surface temperatures in many urban centres may be attributed to the photochemical generation of O_3. There is a need to study the distribution of O_3 and its formation from photochemistry of major pollutants from vehicular exhaust over urban airsheds in India.
- Air quality monitoring as well as modelling studies should focus on quality assurance aspects, minimization of uncertainties and ensuring reliability of the study findings, thus helping various users including policy

makers in effective decision-making and implementation of various air quality management strategies.

- Air quality issues need to be addressed using a holistic approach and carrying out interdisciplinary research with enhanced collaboration between epidemiologists, physiologists and atmospheric scientists to establish long-term measurement programmes to characterize air quality on a regional to global scale.
- It is also necessary to include the following under air monitoring programme:
  (i) Measurements of pollutants, precursors and reactive intermediates;
  (ii) Study of process of formation, evolution, growth and removal of particulates; and
  (iii) Real time particle characterization (Molina *et al.*, 2004).

REFERENCES

Goyal, S.K., Rama Krishna, T.V.B.P.S., Singh, R.N. and Devotta, S., 2005. Indian Air Quality Monitoring and Modelling, Glimpses of the Work on Environment and Development in India, Eds. Singh, J.S. and Sharma, V.P., INSA, Angkor Publishers (P) Ltd., New Delhi, 124-58.

Molina, Mario J. and Molina, Luisa T., 2004. Mega-cities and Atmospheric Pollution, Air and Waste Management Association Annual Meeting, Indianapolis, June 23, 2004.

World Bank, 2002. What do we know about air pollution—India Case Study, Urban Air Pollution—South Asia Urban Air Quality Management Briefing Note No. 4, March 2002.

Government of India (GOI), 2002. Report of the Expert Committee—Auto Fuel Policy, August 2002.

TERI, 2001. Review of past and on-going work on urban air quality in India, report submitted to the World Bank, December 2001.

CPCB, 1994. CPCB—NAAQS Notification dated April 11, 1994.

Air Pollution in Megacities

2

S.K. Goyal

Preamble

Air pollution has become a growing problem in megacities and large urban areas. A megacity is often defined as a metropolitan area with more than 10 million inhabitants. In 1950, New York and London were the only megacities. This number grew to 20 by 2000. Currently, there are more than 100 metropolitan areas with official population exceeding 3 millions (United Nations, 2004).

Exposure to air pollution can result in a number of health problems, from minor illness to death. WHO estimated that every year 800,000 people die prematurely due to lung cancer, cardiovascular and respiratory diseases, caused by outdoor pollution worldwide. Approximately, 150,000 of these deaths are estimated to occur in South Asia alone (World Bank, 2003). Between 100-150 million people around the globe have asthma and the figure is increasing. Over 180,000 people die worldwide each year due to asthma. India has an estimated 15-20 million asthmatics. A huge amount is spent on treatment of diseases caused by air pollution. As per WHO estimates, the US spends more than $ 6 billions on asthma, and Britain spends about US$ 1.8 billions on health care for asthma and because of mandays lost through illness (Anon., 2004).

Current Status

Goyal *et al.* (2005) analysed the air quality status (during 1995-2003) in 19 major cities having population more than one million along with the megacities (Delhi, Mumbai and Kolkata) of the country. The annual mean levels of SPM were in general found to be higher than the CPCB standard of 140 μg/m^3 in all the cities, except at Chennai and Kochi, during all the years. Annual mean levels of SO_2 and NO_2 were found to be always much below the prescribed standard of 60 μg/m^3 during all the years in all the cities. However, Kolkata, Delhi and Bangalore had shown an increasing trend in NO_2 levels after 2001, mainly due to considerable increase in vehicular traffic. PM10 also exceeded the CPCB standard of 60 μg/m^3 in all the cities during 2001-2003. Based on the air quality data, Delhi, Mumbai, Kolkata, Chennai, Kanpur, Ahmedabad, Bangalore and Pune were identified as highly polluted cities, particularly with respect to particulate matter.

International Scenario

Molina and Molina (2004) reviewed air quality status of 20 megacities of the world, wherein detailed data were available for 15 cities. SO_2 and NO_2 data were available for 1998, whereas SPM data were available for 1999. Air quality status along with population in 2003 in these megacities is presented in Fig. 1. Air

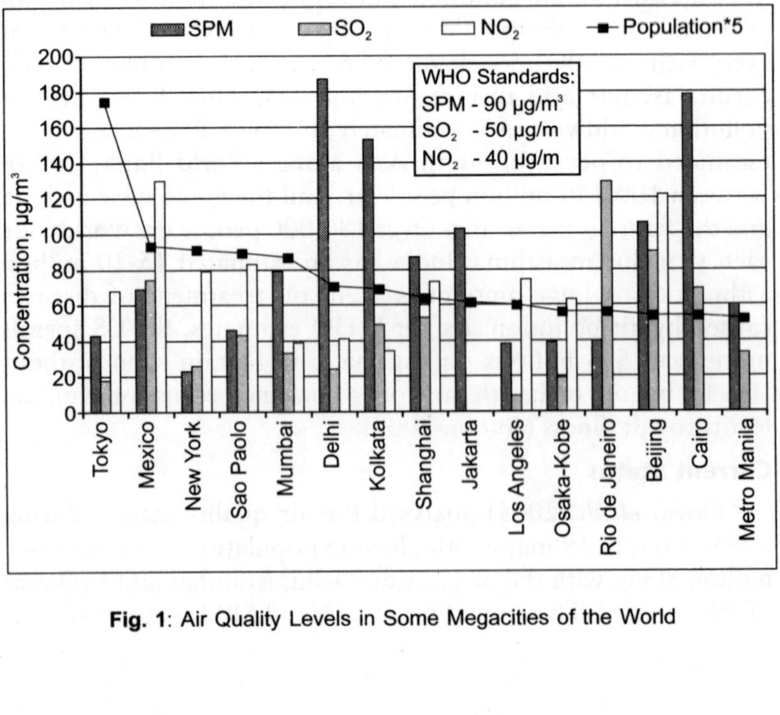

Fig. 1: Air Quality Levels in Some Megacities of the World

quality data were compared with corresponding WHO guidelines. SPM levels were found to be high in Delhi, Jakarta and Cairo. High SO_2 levels were found in Mexico, Rio de Janeiro, Shanghai, Beijing and Cairo. NO_2 levels were found to be high in most of the cities, except in Indian cities including Delhi, Mumbai and Kolkata.

Schwela (2003) analysed the ambient air quality status of 15 major metro-cities in Asia for the period 1990-1999. Particulate matter (SPM and PM10) was found to exceed the WHO standards of 90 and 50 $\mu g/m^3$ respectively for most of the cities. SO_2 levels were found to be below the standards in all the cities, except in Chinese cities: Beijing, Chongqing, Busan and Shanghai. NO_2 levels were found to be either low or around the WHO standard of 40 $\mu g/m^3$.

Most Serious Concerns

Serious concerns in megacities pertain to lack of understanding of sources of air pollution and their inter-linkages with factors such as land use planning, traffic planning, modal transport demand, etc. Though transportation is recognised as the major source of air pollution in many cities, especially in developing countries; sources which are unorganised and responsible for high resuspended dust cannot be ignored. The challenge facing megacities is how to reduce the adverse environmental impacts and other negative effects of transportation without giving up the benefits of mobility. The dilemma becomes most pressing under conditions of rapid urban growth, which is likely to increase travel demand significantly (Molina and Molina, 2002).

Haq *et al.* (2004) assessed the air quality measurement (AQM) capabilities of 20 cities in Asia with an objective to assist the cities in setting priorities and developing strategies for strengthening their AQM capability. A capability index was estimated for each city based on 4 sets of indicators, AQM capacity, data assessment and availability, emission estimates and management enabling capabilities. Out of 20 cities, 6 had excellent, 2 good, 5 moderate, 5 limited and 2 minimal AQM capability.

Quality and quantity of air quality data are the most important concerns in defining the air quality of a city. Several researchers (TERI, 2001; World Bank, 2002; Goyal *et al.*, 2005) emphasised the need for QA/QC in air quality monitoring. Attention is also required in data analysis; timely publication of, and access to, raw data; re-examination of site selection in the light of changes in land-use patterns; the actual monitoring frequency; and the positioning of the instruments at a given site. With limited and of unknown data quality, it becomes often difficult to correlate air quality of a city with various activities and also to evaluate effectiveness of pollution control measures.

Emerging Issues

Analysis of national and international scenario on air quality reveals that at present, particulate matter is the pollutant of concern in most of the countries. However, increasing trend in NO_2 levels are indicated in many Asian cities, viz. Delhi, Kolkata, Pune, Tokyo and Hong Kong, attributed mainly to transport sector. As the number of motor vehicles in megacities (particularly in developing countries) increases, NOx emissions will increase dramatically. Among the criteria pollutants, NO_2 could be the pollutant of concern in urban areas. Fine particulate matter (PM2.5 or even smaller size) could be another pollutant requiring attention due to increased use of gaseous fuels in future.

Inadequacy in Management and Control

Akbar (2004) carried out retrospective analysis of policy interventions and air quality trends (with a focus on particulate pollution) for five metros in India viz., Delhi, Mumbai, Kolkata, Chennai and Hyderabad, over ten years period (1993-2002). SPM and PM10 data collected by NEERI in these cities were analysed. PM10, the main pollutant of public health concern was found to decline, indicating the positive impact of various interventions. However, quality of data was observed to be one of the constraints, and a need for strong framework for monitoring and analysis of interventions was emphasised. Akbar *et al.* (2004) also emphasised that implementation of the same intervention in two cities with different meteorologies and mix of sources is unlikely to give the same results.

Most of the air pollution studies carried out pose problems due to lack of availability of sufficient database and the reliability of available data. For example, the benefits of CNG conversion are yet to be established conclusively in the case of Delhi (Goyal *et al.*, 2005). Guttikunda *et al.* (2004), while reviewing the source apportionment of urban air pollution in Asia and Latin America emphasised that lack of information on the sectoral contributions to air pollution especially fine particulates, is one of the typical constraints faced in formulating an effective integrated urban air quality management programme.

Source apportionment studies conducted without representative source profiles will also hamper devising foolproof management strategies for air pollution control (Kumar, 2004).

Future Challenges

Several researchers (Carter, 2004; Singh, 2002; Dutta *et al.*, 2003; Molina and Molina, 2004; Goyal *et al.*, 2005) reviewed the air quality management issues in various megacities/large urban areas and identified the challenges that need to be addressed to tackle air pollution problem. It was suggested that addressing air quality issues effectively requires a holistic approach: one that takes into account scientific, technical, existing infrastructure, economic, social, and political factors. There is no single strategy for addressing air pollution problems in megacities. A mix of policy measures best suited for each city's challenges and customs will be needed to improve air quality.

An integrated air quality management system will need incorporation of strengthening of institutional mechanism, assessment of air quality (monitoring, emission inventory, source apportionment, air pollution exposure and damage), development of integrated strategy action plans and evaluation of control strategies. The implementation of action plan will need constant check on effectiveness as also mechanism to incorporate changes in the plan midway.

REFERENCES

Goyal, S.K., Rama Krishna, T.V.B.P.S., Singh, R.N. and Devotta, S., 2005. *Indian Air Quality Monitoring and Modelling, Glimpses of the Work on Environment and Development in India.* Editors: J.S. Singh and V.P. Sharma. INSA, Angkor Publishers (P) Ltd., New Delhi, 124-58.

Akbar, Sameer, 2004. Ten Years of Urban Air Quality Management in India—Findings of a Recent Study Across Five Cities, Workshop on Urban Air Quality Management In India: Progress and Future Challenges, New Delhi, Oct. 18-19, 2004.

Akbar, Sameer and Sengupta, B., 2004. Impact of Policy Interventions on Urban Air Quality in India: A Retrospective Analysis, Better Air Quality 2004 Workshop, Agra, Dec. 6-8, 2004. Abstracts Compendium, 13.

Anon., 2004. A Disease on the Rise. Down to Earth, March 15, 33-34.

Carter Michael, 2004. Air Quality Management in Asia. Better Air Quality 2004 Workshop, Agra, December 6-8, 2004.

Dutta, S.A., Sengupta, B. and Biswas Dilip, 2004. Strategic Needs for Integrated Air Quality Management in India, Workshop on Urban Air Quality Management organized by the CPCB and World Bank, New Delhi, October 18-19, 2004.

Gray, Haq, Dieter Schwela, Cornie Huizenga, Herbert Fabian and May Ajero, 2004. Bench-marking Urban Air Quality Management in 20 Asian Cities: Initial Final Analysis, Better Air Quality 2004 Workshop, Agra, December 6-8, 2004, Abstracts Compendium, 8.

Guttikunda, S., Johnson, T., Russell, A. and Procee, P., 2004. Source Apportionment of Urban Air Pollution in Asia and Latin America, Better Air Quality 2004 Workshop, Agra, December 6-8, 2004, Abstracts Compendium, 9.

Kumar, R., 2004. Review and Results of Studies on Inventories and Source Apportionment, Better Air Quality 2004 Workshop, Agra, December 6-8, 2004.

Molina, Mario J. and Molina, Luisa, T.,2004. Mega-cities and Atmospheric Pollution, *Journal of Air Waste Management Association*, Vol. 54, 644-80, June 2004.

United Nations, 2004. Population Division, World Urbanization Prospects: The 2003 Revision. (http://www.un.org/esa/population/publications/wup2003/2003WUP Highlights.pdf).

Schwela, D., 2003. Strategic Framework for Air Quality Management in the Asian Region, BAQ 2003, Manila, Philippines, December 17-19, 2003.

World Bank, 2003. Health Impacts of Outdoor Air Pollution, in Urban Air Pollution—South Asia Urban Air Quality Management Briefing Note, No. 11, February.

Molina, L.T. and Molina, M.J., 2002. *Air Quality in the Mexico Megacity: An Integrated Assessment*. Eds. Molina, L.T. and Molina, M.J.; Kluwer Academic Publishers: Dordrecht, The Netherlands.

Singh, R.N., 2002. Air Quality Management in Indian Cities: Trends and Challenges, Better Air Quality in Asian and Pacific Rim Cities (BAQ-2002), December 16-18, Hong Kong.

World Bank, 2002. What do we know about air pollution—India Case Study, in Urban Air Pollution—South Asia Urban Air Quality Management Briefing Note, No. 4, March.

TERI, 2001. Review of Past and On-going Work on Urban Air Quality in India, report submitted to the World Bank, December.

Urban Air Pollution in Delhi and Mumbai—Issues and Challenges

3

Rakesh Kumar

Preamble

Deteriorating urban air quality is a serious concern for human health and welfare in most of the large cities across the globe; however, this problem has not been adequately addressed in many cities in developing as well as South Asian cities. These cities have seen rapid economic development which is likely to further dip the graph for better air quality goals. Though vehicular and industrial pollution have been seen as the leading contributors to urban air pollution in many cities, ground level anthropogenic activities such as refuse burning, uncontrolled biomass burning, resuspended dusts, etc. have created local hotspots. However, most of the air quality clean up programmes have mainly focused on emissions from industrial and mobile sources.

The cause of concern in recent times has also shifted to the assessment of effect of fine particulates on human health (Pope *et al.*, 1995) in comparison to other commonly measured air pollutants such as total particulate matter, sulfur dioxide, nitrogen oxides, carbon monoxide, lead and in some cases, ozone. Most of the cities in Asia have eliminated lead additives in gasoline. India has phased out lead completely in the year 2000. Based on the air quality health impacts being witnessed worldwide as also in India (Balakrishnan *et al.*, 2003; Chowgule *et al.*, 1998), it has become imperative that the health cost of air pollution has become too high to ignore and, therefore, there is a need to look at the strategies for air pollution control in Indian cities.

Current Status

Though many cities in India are plagued with poor air quality, major cities such as Delhi and Mumbai have attracted more attention as these two cities are considered political and economic capitals, respectively for India. With a population of 14 millions (Census 2001), Delhi ranks among the largest cities of the world whereas Mumbai is not too far behind at 12 millions. Delhi, the capital of India, stands at the western end of the Gangetic Plain, has an arid climate most of the year. It gets high contributions from wind-blown dust. Other major contributing sectors in Delhi are transport and three power plants located nearby. On the other hand, Mumbai is a coastal city, with three sides surrounded by sea, harbour and Thane creek, respectively. Its climate is highly influenced by sea-land breeze phenomena along with four months extended monsoon. The city witnesses contribution of air pollution from traffic, industries (it has shown declining trend), resuspended dusts due to large scale construction/demolition and refuse burning (PM Reduction Action Plan Report, NEERI, 2003, Kumar *et al.*, 2003). Emission inventories prepared in some cities of India are based on emission factors of USEPA : AP 42 (Air-Chief, 2004) and some limited data sources available in India. Action plans as well as interpretation based on these emission factors have led to misleading conclusions in the past.

International Concerns

Urban air quality deterioration has been not only responsible for declining quality of life but also for global air pollution problem due to trans-boundary movements impacting outside political boundaries as well. One of the major features, which is of international concern, is health. What also worries internationally about these urban centers is also the pattern of growth being witnessed in these cities. Some of the common features, which are common to cities in India as well in neighbourhood and South Asia are, presence of large numbers of two-stroke engine vehicles; aged vehicle fleet; unregulated small scale industries in residential areas; inadequate air quality equipment; lack of training and resources; unreliable air quality monitoring data; lack of emission factors and emission inventories.

For delineating action plans, one also needs to have better understanding of impact of air pollution on health and welfare (HEI, 2003). A 1995 World Bank study estimated that the annual health costs of ambient air pollution in Delhi were some 3.5–14 billion rupees, whereas an another study (Kumar *et al.*, 2002) for Mumbai, indicated a loss of about Rs. 11,224 millions.

Most Serious Concerns

In India, various air quality improvement action programmes have been implemented, particularly in Delhi and Mumbai, where series of Supreme Court and High Court orders have been passed, viz., introduction of CNG for commercial vehicles in Delhi, ban on sale of loose lubricants for 2-stroke vehicles, sulphur and benzene reduction, mandating of higher version of EURO II and EURO III emission standards, ban on commercial vehicles (more than 8 years old) plying in Mumbai, PUC upgradation etc. Even though these actions have led to immediate improvement in air quality in these two cities, the current rate of increase of vehicles on road is going to negate the benefits generated through various measures due to high congestion resulting in low vehicle kilometers travelled per unit time as well as fuel. Vehicular pollution control seems to be an attractive option but some recent studies on source apportionment have indicated that the air pollution contribution is not just limited to vehicular sector alone (Kumar *et al.*, 2003). Lack of understanding of sources, their transport mechanisms and contribution to air quality at receptors are the major issues in developing the strategies for meeting better air quality goals.

Urban air shed is not limited to the political boundary of the cities like Delhi or Mumbai but most of the management solutions are devised to address these limited political boundaries. Growth of smaller cities outside Delhi periphery as also large scale growth of distant suburbs in Mumbai would drive the air quality goals in future. Serious concerns in both these cities are with regard to larges scale migration within the city as also exponential growth of urban centers just beyond the boundary of the megacity.

Emerging Issues

Air pollution control targets in recent times may have shifted from the conventional pollutants such as total particulate matter, sulfur dioxide, nitrogen oxides, carbon monoxide, lead and ozone to fine particulate matter, however, the potential of other pollutants and their combined effect on the human health can not be ignored in the long run. One pollutant of concern, which will emerge in medium and long-term as a major pollutant, can be NOx. Though vehicular sector shall make an attempt to meet the NOx and HC emission requirements as mandated by EURO III and EURO IV, the overall addition to number of vehicles on road will lead to high levels of ambient ozone levels, which till date is much lower. [Daily average ozone readings of about 30 $\mu g/m^3$ recorded in Delhi are far below the US 8-hour by average of 0.08 ppm (equivalent to 157 $\mu g/m^3$) (http://www.rrcap.unep.org/issues/air/maledec/baseline/Baseline/India/INCH2.htm).] Addition of air emissions from power plants and other activities could add to the problem. Suspended Particulate Matter values are high, particularly in winter season (in Delhi, it exceeds by two times the national standards for PM10, whereas in Mumbai it exceeds marginally for limited period). Therefore, in the long run, India would experience the problems of increased NOx and PM10 ambient values. It would also lead to high levels of ground level ozone values.

Inadequacy in Management and Control

Many initiatives taken by the Government as well as the Court's directives have led to improvement in the air quality. However, the driver for better air quality has not been scientifically assessed and implemented. Some of the inadequacies can be linked to the fact that conversion of public transport in Delhi to CNG alone is responsible for the overall improvement in air quality. Better air quality goals have not been previously defined under one umbrella till autofuel policy document was prepared by Dr. Mashelkar's Committee [petroleum.nic.in/afp_con.htm]. This committee outlined a roadmap for implementation describing the roles of various departments, agencies and sectors. It also brought out the issue of lack of adequate knowledge and information to initiate actions. Another issue relating to management and control pertains to multiple role of many agencies for taking required steps. Formation of EPCA by the Supreme Court can be considered as a partial effort in this direction for the cities cited by the Supreme Court. For other cities, multiplicities of roles are yet to be resolved. Megacities' problems may be similar in case of population explosion, infrastructure for water and wastewater treatment as also sanitation, however, air pollution problem could be unique and needs management approaches which integrate the local expectations, fit in the overall developmental schemes of the city and meet the requirement of better quality of life.

Future Challenges

First and foremost challenge for air pollution abatement in India relates to improving the current state of knowledge on sources, emission factors for these sources, preparation of dynamic inventories for constant upgradation (Biniwale *et al.*, 2005) and source apportionment for strategic development of action plans. Fuel mix and fuel adulteration (CPCB, 2003) are the major drivers of air pollution scenario in cities and therefore effective tools to manage them are warranted. There is a need to develop city specific solutions based on the local sources, emission characteristics, meteorological conditions, institutional support and infrastructure. Also long range transport of pollutants, particularly particulate matter, needs to be better understood in tropical regions, where large scale background concentration may be contributed by outside sources.

R&D challenges in this area include targeted assessment to provide results for taking ground level actions for better local and regional air quality. All the efforts need to be linked to health improvement with proper assessment of benefits to the masses using economic principles for communicating the successes and failures.

REFERENCES

Biniwale, R.B. and Kumar, R., 2005. Fuel Based Emission Inventory for Planning Gasoline Quality Modifications in Indian Mega city, *Society of Automotive Engineers Journal.*

Balakrishnan, K., Ganguli, B., Padmavathi, R., Sankar, R., Sankar, S. and Thanasekaraan, V., 2004. Estimating Health Impacts of Outdoor Air Pollutants using Exposure—Response Functions from Time-series Analyses in Chennai, CAI-Asia Conference, Better Air Quality 2004, Agra.

NEERI, 2004. Particulate Matter Reduction Action Plan for Greater Mumbai, 2004, MMRDA Report prepared by NEERI.

CPCB, 2003. Transport Fuel Adulteration, *Parivesh*, July 2003.

Kumar, R., Biniwale, R., Rao, C.V.C., Tondwalkar, V. and Phatak, V.K., 2003. Inventorization and Source Apportionment of Particulate Matter Air Pollution in Mumbai, India, proceedings of CAI-ASIA organized BAQ 2003, Manila, December 17-19, 2003.

HEI (Health Effects Institute), 2002. HEI Perspectives : Understanding the Health Effects of Components of the Particulate Matter Mix: Progress and Next Steps. (http://www.healtheffects.org/Pubs/Perspectives-2.pdf)

Kumar, R. and Srivastava, A., 2002. Economic Evaluation of Health Impacts of Air Pollution in Mumbai, Environmental Monitoring and Assessment, Vol. 75, 135-43.

Jacobson, M.C., Hansson, H.C., Noone, K.J. and Charleson, R.J., 2000. Organic Atmospheric Aerosols: Review and State of Science, Reviews of Geo-physics, 38, pp. 267-94.

Chowgule, R.V., Shetye, V.M., Parmar, J.R., Bhosle, A.M., Khandagale, M.R., Phalnitkar, S.V. and Gupta, P.C., 1998. Prevalence of Respiratory Symptoms, Bronchial Hyperactivity, and Asthma in a Megacity (Mumbai), *American Journal of Respiratory and Critical Care Medicine*, 158, pp. 547-54.

World Bank, 1995. Brandon, C. and K. Homman. The Cost of Inaction: Valuing the Economy-wide Cost of Environmental Degradation in India.

Pope, C.A., Dockery, D.W. and Schwartz, J., 1995. Review of Epidemiological Evidence of Health Effects of Particulate Air Pollution, Inhalation Toxico, 7: 1-18.

UNEP, CPCB data. (http://www.rrcap.unep.org/issues/air/maledec/baseline/Baseline/India/INCH2.htm)

Air Pollution from Industries 4

S.M. Tamhane

Preamble

Thermal Power Plants, Cement Plants, Petroleum Refineries, Steel Plants, Non-Ferrous Smelters, Fertilizers and Fine Chemicals plants, etc. are the major industrial sources with significantly high air pollution potential. Most of these large/medium scale industries are energy intensive and consume millions of tonnes of solid/ liquid/gaseous fuels and emit considerable quantity of emissions of criteria pollutants (SO_2, NOx, CO, HC, PM10) unless, they are effectively controlled. Apart from the criteria pollutants, formation of persistent organic pollutants (POPs) and persistent bio-accumulative toxic chemicals (PBTs) particularly in high temperature combustion processes using alternate/secondary fuels, is the cause of prime concern to national policy makers as POPs and PBTs are toxic to human as well as wildlife and have the tendency to accumulate in the tissues of most of the living organisms and move up in food chain.

Thermal Power Plants

Total estimated coal resources of India are around 205 billion tonnes which is less than 1% of world reserves and 80% of which are of inferior grades with ash content varying between 25-45% (Sachdev *et al.*, 1998). The disadvantages of using high ash content and inconsistent quality coals are high transportation cost, wear and tear of coal handling equipment, reduced flame stability, slogging and fouling of water wall, reduced plant performance, high emission factors and related environmental issues, etc. In view of the above, MoEF has recently mandated the use of beneficiated coal having less than 34% ash content in all power plants located more than 1000 kms. away from coal sources and in urban/environmentally sensitive locations (TERI) (http://www.teriin.org/opet/reports.htm).

India's per capita energy consumption has rapidly increased from 4.16 quadrillion BTU (quads) in 1980 to 12.8 quads in 2001, higher than South Korea (8.0 quads) and Indonesia (4.0 quads) but less than the developed countries like Germany (14.35 quads), Japan (21.92 quads), China (39.67 quads) and the United States (97.05 quads). As regards fuel-wise/energy resource-wise contribution towards energy generation/consumption, coal accounts for 50% (6.5 quads), while petroleum 34.4% (4.4 quads), natural gas 6.5%, hydroelectricity 6.3%, nuclear 1.7% and geothermal, wind, solar, biomass together account for the remainder (DOE, 2004).

Power Plant Emissions

In Asia, India's carbon intensity [Carbon emission MT/$ (1995) of GDP] in 2001 is 500 MT Carbon/$, next only to China with the highest value of 750 MT/$ and is relatively high in comparison to other neighbouring countries. Indian economic policies such as high import duty on high quality coal and subsidies on low quality domestic coal explain the reasons for high reliance on and use of low quality coal which resulted in country's high carbon intensity levels (DOE, 2004). According to an estimate (Ohio Supercomputer Centre) (http://www.osc.edu/research/user/index.shtml), total CO_2 emissions from all the coal fired power plants in India were 1100 tons/day in 1997-1998. Computed values of CO_2 emissions per unit of electricity generated range between 0.8 and 1.2 kg/kWh. Total SO_2 emissions from all the power plants in India have been estimated to be 7600 tons/day. Average SO_2 emissions per unit of electricity are found to be 7.4 gm/kWh. NOx emissions from TPPs are 7900 tons/day. Most of the plants have NOx emissions in the range of 6-10 gm/kWh. SPM emissions vary from 2.5-3.5 gm/kWh of electricity generated. Introduction of government's new initiatives to discourage firms generating power using old inefficient plants, reduction in subsidies on using old low quality coals and fiscal initiatives such as soft loans for installation of pollution control equipment, customs waiver for high quality coal etc. would go a long way in bringing down the carbon intensity levels in India.

Cement Industry

The cement manufacturing process involves mining, crushing, grinding of limestone, clay, coal, calcining raw materials in rotary kilns, cooling the clinker, mixing with gypsum, milling and finally bagging the product. The entire process is energy intensive and generates dust, most of which is collected and recycled partially to the process. The dry process with energy consumption of around 200 J/kg, almost half that of wet process, is a techno-economically as well as environmentally preferred and worldwide accepted process. Well designed, operated and maintained plant emits around 0.2 kg dust/MT clinker *vis-à-vis* NOx emissions of 1.5 kg/MT of clinker for dry process and 4.5 kg/MT for wet process plants. SO_2 emissions due to oxidation of sulfur in the fuel can be eliminated to the tune of 90% because of highly alkaline conditions prevailing in the kiln. For cement plant, operating under full load conditions, source emission guidelines (World Bank, 2003) acceptable to World Bank Group for financial assistance are as follows: Particulates –50 mg/Nm^3 equivalent to 0.2 kg/ton of clinker, NOx–600 mg/Nm^3, SO_2–400 mg/Nm^3.

Iron and Steel Industry

Although iron and steel industry is an energy intensive industry because of recyclability of scrap off cuts, spilled steel rejects, etc., the industry makes significant contribution towards energy and materials conservation and waste minimization. Carbon emissions arise both from energy use and from certain industrial processes involved in the production of iron and steel. In general, the integrated steel industry releases around 2.0 to 2.5 tonnes of CO_2 per ton of steel (Fletcher, 2005; NEERI, 2000). During the last 40 years, the energy consumption of steel making industry has gone down by 60% while the total CO_2 emissions of the steel industry have been reduced by 50% over the same period. The carbon intensity can be further reduced with appropriate changes in the fuel mix and carbon content of the fuels, introduction of technology based on recycling of blast furnace top gas after de-carbonization, CO_2 capture and storage technologies, etc. (Fletcher, 2005). Emissions of criteria pollutants viz. particulates, SO_2 and NOx are respectively in the range of 0.35 to 5.3, 0.03 to 1.34 and 0.02 to 1.7 kg/MT from various units in the Indian steel plants (NEERI, 2000; NEERI Report, 1987). While reported particulate emissions in United States are much higher in the range of 10 to 40 kg/MT of steel, SO_2 and NOx emissions from similar units in India are comparable (Cherimisinoff, 2002).

Refineries

There are 17 oil refineries presently operating in India having an annual installed capacity of 113 MMTPA likely to be enhanced to 179 MMTPA by 2006 while the projected demand by 2025 is 370 MMTPA (Bansal, 2002). Number of air pollutants are released during combustion of various fuels as well as during other oil refining operations. Control of total organic emissions (including VOCs) to a level of 0.6% of the throughput, can be considered as a good target for a petrochemical complex. Emissions from refineries account for 6-7% of the total emissions resulting from the production and consumption of petroleum products in the country. Various units of a typical Indian refinery, in total, emit around 3.62 kg SO_2, 0.134 kg SO_3 and 1.435 kg NOx per MT of feed (NEERI, 1987). Various facilities required to be installed for reducing the refinery emissions levels are: (a) hydro-desulphurization of feed, (b) flue gas desulphurization, (c) tail gas treatment, (d) provision of ESPs, (e) additional stages of cyclones in Fluidized Catalytic Cracking Units (FCCU), (f) provision of additional conversion stages in SRU from sulphur removal, etc.

Fertilizer Plants

Particulates from urea plant (UP) and rock grinding plants (RGP), fluorides from di-ammonium phosphate (DAP) and phosphoric acid plant (PAP), SO_2 from sulphuric acid plant (SAP) and NH_3 from DAP and UP are the major pollutants released from a fertilizer plant. Emissions from a typical Indian fertilizer plant include particulates: 0.5-1.9 kg/MT from UP and 0.75 kg/MT from RGP, Fluoride: 2 kg/MT each from DAP and

PAP, NH_3: 2 kg/MT from DAP and 0.26 - 0.65 kg/MT from UP, and SO_2: 1.6-1.8 kg/MT from SAP.

Green House Gas (GHG) Emissions

The national GHG inventory prepared by NPL (CSIR), Delhi estimates total carbon dioxide (CO_2) equivalent (comprising CO_2, CH_4 & N_2O) emissions to be 1,001,352 Gg which is around 3% of total global CO_2 equivalent emissions (TERI, 2003). The distribution of sectoral contribution includes 57% from the energy sector (excluding biomass burning), 34% from agricultural sources, 7% from waste sector, 2% from industrial processes and

Table 1. CPCB Industrial Emission Standards

Sl. No.	Type of Industry	Unit	Capacity/ Operation Mode	Emission Standard/ Stack Height (m)	EPA Notification Date
1.	Thermal Power Plant	Coal fired boiler	+500 MW +210 MW -210 MW	PM = 150 mg/Nm^3, H=275m PM = 150 mg/Nm^3, H=220m PM = 300 mg/Nm^3, H= $14Q^{0.3}$	GSR 742 (E) (Aug. 30, 1990) GSR 7
		Naptha fired boiler	-100+400 MW	PM = 100 PPM	(Dec. 22,1998)
		NG fired boiler	-100 MW +100-400 MW +400 MW	NOx = 100 PPM NOx = 75 PPM NOx = 50 PPM	GSR 7 (Dec. 22,1998)
2.	Integrated Iron & Steel	Sinter plant		PM = 150 mg/Nm^3	SO 64 (E) (Jan. 18, 1988)
		Steel making	Normal Oxygen lancing	PM = 150 mg/Nm^3 PM = 400 mg/Nm^3	
		Rolling mill		PM = 150 mg/Nm^3	
		Refractory material		PM = 150 mg/Nm^3	
		Coke ovens		PM = 50 mg/Nm^3 CO = 3 kg/ton Coke	
3.	Oil Refinery	Distillation	(Atm.+Vacuum)	SO_2: 0.25 kg/ton feed	GSR 742 (E) (Aug. 30, 1990)
		Catalytic Cracker Unit		SO_2: 2.5 kg/ton feed	
		Sulphur Recovery Unit		SO_2: 120 kg/ton "S" in feed	
4.	Fertilizers & Fine Chemicals	Prilling Tower (Urea Plant)	Comd. Prior dt. 1-1-1982	PM = 150 mg/Nm^3 or 2.0 kg/ton	SO 63(E) (Jan. 18, 1988)
			After	PM = 50 mg/Nm^3 or 0.5 kg/ton	
		Phosphoric Acid Plant		F = 25 mg/Nm^3 PM: 150 mg/Nm^3	
5.	Pulp & Paper Mill			PM = 150 mg/Nm^3 H_2S = 10 mg/Nm^3	GSR 913 (E) (Oct. 24, 1989)
6.	Sulphuric Acid Plant			SO_2: 4 kg/MT(100%) H_2SO_4 Acid mist: 50 mg/Nm^3	SO 64 (E) (Jan. 18, 1988)

MW – Megawatt, PM – Particulate Matter, SO_2 – Sulphur Dioxide, CO – Carbon Monoxide, NOx – Oxides of Nitrogen, F – Fluoride, H – Height of stack, Q – SO_2 emission rate, kg/hr.

almost negligible from land use change and forestry sector. Although India figures in amongst top 10 contributors of GHG emissions, its per capita share is relatively low. Based on 1990 estimates, per capita CO_2 equivalent emissions work out to be 1.194 tonnes or 325 kg carbon which is one-sixth of the world average (Asian Development Bank, 1994) but far less than per capita emissions of Japan: 2400 kg carbon and U.S.: 5400 kg carbon (ALGAS—India, 1998).

Summary of MoEF approved CPCB Environmental (Industrial Emission) Standards (CPCB Publication, 2001) is presented in Table 1.

Recommended Measures for Cost-Effective Air Pollution Management

- Latest technology based energy efficient machines/equipment need to be deployed for energy generation using cleaner fuels with low sulfur and ash content or refined fuels viz. beneficiated coal, desulphurised fuel oil or fuel mix.
- Low NOx burners with optimum air/fuel ratio need to be used and wherever possible, flue gas heat recovery systems be installed to optimize fuel consumption, load on control equipment and reduce emissions.
- Tail/tertiary gas treatment with regenerative environment friendly processes need to be adopted with maximum recycle and minimum purge systems.
- Cost-benefit analysis for environmental protection should be made through trade-off between capital and recurring (annualized) investments required for replacing with modern technologies/equipment *vis-à-vis* retrofitting existing equipment.
- In the event of likely cumulative impact of national as well as international GHG emissions on temperature and sea level rise, water and land management strategies and land use plans for coastal areas are to be appropriately restructured.
- New processes need to be developed based on break-through technologies for achieving post Kyoto protocol

requirements, in order to have sustainable industrial development.

REFERENCES

Anthony, F., 2005. Steel Industry Resolves to Achieve Long-term CO_2 Reductions. (http://www.foodproductiondaily.com/news)

DOE, 2004. India: Environmental Issues, Energy Information Administration. (http://www.eia.doe.gov/emeu/cabs/indiaenv.html)

World Bank, 2003. Pollution Prevention and Abatement for Cement Industry. (http://www.worldbank.org)

TERI, 2003. India's Contribution to Greenhouse Gas Emissions. (http://www.teriin.org/about/abteri.htm)

Bansal, K.M., 2002. Transportation of Crude and Products In and Out of Indian Refineries, Indian Chemical Engineer, Vol. 44, No. 1, Special Issue on Petroleum Refining in India, October-December, 2002.

Cherimisinoff, N.P., 2002. Handbook of Air Pollution Prevention and Control-Cherimisinoff, N.P., 2002, Elsevier Publication, First Indian Reprint, 2004, 121.

CPCB, 2001. Environmental (Industrial Emission) Standards, July 2001.

NEERI, 2000. Life Cycle Assessment of Steel Sector in India, May 2000.

Sachdev, R.K., 1998. Beneficiation of Coal—An Economic Option for the Indian Power Industry, World Energy Congress on Energy and Technology: Sustaining World Development Into The Next Millennium, HOUSTON, USA, September 1998.

ALGAS—India, 1998. (Asia Least-Cost Greenhouse Gas Abatement Strategy) Manila: Asian Development Bank, 238.

Asian Development Bank, 1994. Climate Change in Asia: India Country Report, Manila, 172.

NEERI, 1987. Impact of Stack and Fugitive Emissions from Bhilai Steel Plant on Neighbourhood Air Quality and Evaluation of Air Pollution Control System, August 1987.

NEERI, 1987. Impact of Fugitive and Stack Emissions from Selected Industries on Neighbourhood Air Quality (Vol. III: Petroleum Refinery).

Ohio Supercomputer Centre, Anthropogenic Emissions from Energy Activities in India: Generation and Source Characterization. (http://www.osc.edu/research/user/index.shtml)

TERI, Technology Status of Thermal Power Plants in India and Opportunities in Renovation and Modernization, an OPET International Action on Refurbishment of Thermal Power Plants in India. (http://www.teriin.org/opet/reports.htm)

Air Pollution from Small and Medium Enterprises

5

Padma S. Rao

Preamble

Small and Medium Enterprises (SMEs) are playing an important role in the Indian economy as they are cost effective as well as labour intensive, thus creating job opportunities and providing low investment per worker. They help in rural development and decentralisation of economic power. In addition, small units support entrepreneurial talent and skills, stimulate personal savings, and help in developing innovative and appropriate indigenous technology providing dynamism and contributing to competition (Rajendran, 1989).

Current Status

SMEs are defined as those with less than US$ 180,000 capital equipment (US-AEP, 1996). In India, the air polluting SMEs include units like kilns, crushers, foundries, small boilers, glass and ceramic industry, etc. SMEs are one of the major employers and provide goods and services to large enterprises. According to SIDBI report on small scale industries sector, 1999, there are more than 3 million SMEs (units) employing nearly 16.7 million persons, and account for 35% of India's total exports and about 40% of industrial manufacturing. In real terms, the small-scale sector recorded a growth rate of 10.1% in 1994-95 as against 7.1% in 1993-94 and 5.6% in 1992-93. By the year 2025, if not regulated, this sector will grow even more rapidly (Parthasarathy, 1996). The SMEs in India are mostly unorganized, uncontrolled and are found in clusters. They use primitive technology, low grade fuel and inefficient systems and thus generate enormous

Table 1. Air Pollution from Small/Medium Enterprises

Sl. No.	Small/Medium Scale Industries	Flow Range (m^3/hr)	Gaseous Pollutants (mg/Nm^3)			Dust Pollution (mg/Nm^3)	
			SO_2	NOx	VOC	Conc.	Particle size & its %
1.	Lime Kiln	1000 - 2000	0.24 - 0.42	2 - 4	300	1000 - 2000	80 - 90% are < 10μ size
2.	Foundry	1500 - 2000	230 - 300	23 - 76	300 - 500	800 - 900	80 - 90% are < 15μ size
3.	Brick Kiln	2500 - 4000	0.26 - 1.12	0.08 - 0.2	1000	300 - 2000	50 - 60% are < 10μ size
4.	Stone crusher	—	—	—	—	> 10000	50 - 60% are > 15μ size
5.	Bagasse fired boiler	1500 - 5000	—	—	400 - 500	400 - 1500	80 - 90% are < 10μ size
6.	Rice husk boiler	800 - 3500	—	—	300 - 500	300 - 2000	80 - 90% are < 9μ size
7.	Glass Industry	1500 - 4000	200	500	300	2000 - 4000	80 - 90% are < 10μ size
8.	Small Boilers (coal fired)	800 - 1300	300	800	500	900 - 2000	80 - 90% are < 15μ size
9.	Small Boilers (oil fired)	800 - 1000	100	900	200	100 - 300	80 - 90% are < 10μ size

air pollution which mostly includes dust, VOCs and acidic gases in excess of the regulatory limits. This has resulted in various occupational health problems among the exposed population.

During the last decade, CPCB and SPCBs provided guidelines for controlling air pollution from these industries. These regulatory agencies along with various R&D organizations like NEERI, TERI, NPC and CBRI, etc. undertook various emission measurements and developed systems for control of pollution or new improved technologies. But the air pollution from all the sources is still exceeding the standards as the SMEs have not set up adequate air pollution control systems in them. This may be due to shortage of capital, limited access to technology, added burden of using obsolete equipment, lack of skilled manpower, limited access to information on pollution prevention and control systems.

Most Serious Concerns

Small industries, in comparison to large industries lack environmental commitment, technical expertise in environmental management, and the financial capabilities to address environmental problems. Nor do they have standards or effective treatment opportunities and services (Nyati, 1988). They also lack additional space for installing pollution control facilities. There are difficulties in obtaining the technical assistance of reputed and knowledgeable consultants. Since most of the units are scattered, they find it difficult to come together for a joint or common treatment plant. The concern about depressed profit margins and decline in competitiveness prevents these units from using pollution control measures. More emphasis is laid on new investments, production, and other return oriented opportunities. Soft loans for pollution control measures are not lucrative. There are subsidies offered for investments in pollution control as incentives, but the impact of these incentives on these units is little or nothing, for they do not alter the cost-benefit analysis in favour of pollution control investments (Nyati, 1988).

Inadequacy in Management and Control

Although Indian courts closed almost 1,000 factories due to pollution problems, and the Supreme Court fined 15 plants including some multinationals (Johannson, 1994; Shaman, 1996), the effectiveness of these regulatory pressures and compliance is yet to be realized. There is a need (NEERI NCR Report, 1995) to review the issue of air pollution and address its management strategies on short as well as long-term basis. The short-term air pollution mitigation options would include identification of areas of immediate concerns in such SMEs like end-of-pipe treatment and control at source-receptor pathway while the long-term mitigation strategy would include use of cleaner fuels, efficient combustion systems, improved technology, etc.

The source control strategies include use of dust control systems like settling chambers, cyclones, multi cyclones, scrubber and bag filter and water spray for dust suppression while the strategy for control at source receptor pathway would include development of high canopy (foliage area) green belt with pollution dust tolerant species around the industrial units.

International Scenario

In many South Asian and Latin American countries and China, the status of air pollution from SMEs is of similar nature. Also in developed countries the large scale enterprises are increasingly down sizing and depending on network of SMEs. Industrialization, although seen as a solution for promoting economic growth, demands resources (inputs) and releases discharges (outputs) in the form of pollutants. However, the sustainable development warrants industry (Welford and Bhargava, 1996; Hart, 1997) to produce high levels of output using lower levels of inputs and generate less wastes with more effective use of raw materials in production that would eventually result in diminishing costs. Hence, sustainable development should be made the core objective governing the operations of small industries.

Future Challenges

Since in a country like India which experiences perennial population problem, with today's current clean technology and strategic management systems (CSIR Networking Projects 2004-07, TERIs CDM project), SMEs can be effective in reducing the gravity of air pollution impacts. However, regulatory pressure and compliance have to be made stricter so that SMEs can take more proactive and strategic approach to air pollution management (Schoemaker and Schoemaker, 1995) than spending more time in fighting regulations. Hence, the pressure for sustainable vision in these SMEs lies within themselves. They must realize the importance of environmental quality management that could be highly effective if it is administered by the small units themselves. CPCB has recently published guidelines (CREP regulation) for seventeen major polluting industries wherein the entrepreneurs have to proactively manage the pollution. A similar approach may be followed for SMEs or their associations.

REFERENCES

CPCB, 2003-04. CREP Regulations for 17 highly polluting category of Industries.

CSIR Networking Projects, 2004-07. Planning Commission, Govt. of India.

Rao, P.S., Rao, B.S., Manthapurwar, N.S. and Hasan, M.Z., 2003. Flue Gas Treatability Studies. A Tool for Techno-Economic Control of Industrial Air Pollution. Environmental Monitoring and Assessment, 82, 75-81.

TERI, 2003. Report on Identification of Small-scale Industrial Clusters for CDM (Clean Development Mechanism) Projects—a Baseline Study, Report No. 2002IE45, prepared for New Energy and Industrial Technology Development Organisation (NEDO).

CBRI, Roorkee, 2001. Pollution Control System for Lime Kiln—Technology Offers from CBRI, Roorkee.

D'Souza, C.M., 2001. Integrating Environmental Management in Small Industries of India, *Electronic Green Journal*.

CPCB and UPPCB Report, 2001. Report of the Joint Inspection Team of CPCB and UPPCB in terms of the Order of Hon'ble Supreme Court, dated 12-12-2001 in I.A. No. 230-250/2001 of W.P. (Civil) No. 13381/1984, M.C. Mehta *Vs* Union of India and Ors., Chapter XI, Prosecutions Launched, Convictions Secured and Directions given for Closure of Polluting Industries.

CPCB and MoEF, July 2000. Environmental Standards for Ambient Air, Automobiles, Fuels, Industries and Noise.

Vedantam, V., 2000. India's Billionth, The Christian Century, 117(26), 945.

Wahlstorm, M., Laine-ylijoki, J., Maattanen, A., Luotojarvi, T. and Kivekas, L., 2000. Environmental Quality Assurance System for use of Crushed

Mineral Demolition Wasted in Road Constructions. Waste Management, 20(2). 225-32.

SIDBI, 1999. Report on Small Scale Industries Sector, Lucknow, India: Small Industries Development Bank of India.

Dasgupta, S., Laplante, B. and Mamingi, N., 1998. Capital Market Responses to Environmental Performance in Developing Countries (World Bank Policy Research Working Paper # 1909). (http://www.worldbank.org/nipr/work-paper/index.htm)

Hart, S., 1997. Beyond Greening: Strategies for a Sustainable World, *Harvard Business Review*, 67-76.

Lau, R. and Srinivasan, R., 1997. Strategic Issues of Environmental Management, *South Dakota Business Review*; 56(2), 1, 4.

NEERI, 1997. Report on Assessment of Dust Emissions from Stone Crushing Industry in Tamil Nadu, Sponsor Tamil Nadu Pollution Control Board.

Pargal, S., Mani, M. and Huq, M., 1997. Inspections and emissions in India: Puzzling survey evidence on industrial water pollution (World Bank Policy Research Working-paper # 1810). (http://www.worldbank.org/nipr/work-paper/index.htm)

Gupta, M. and Sharma, K., 1996. Environmental Operations Management: An Opportunity for Improvement, *Production and Inventory Management Journal*, 40-46.

Mani, M., Pargal, S. and Huq, M., 1996. Does Environmental Regulation Matter? Determinants of the Location of new Manufacturing Plants in India in 1994 (World Bank Policy Research Working Paper # 1718). (http://www.worldbank.org/nipr/work-paper/index.htm)

Mithra, R. and Richaria, L.K., 1996. Pollution Potential in Lime Stone Aquifers, *Indian Journal of Environmental Health*, 38(4), 256-60.

Parthasarathy, S., 1996. Whither Small Scale Industry? In The Hindu Survey of Indian Industry, Madras, India: Kasturi & Sons, National Press, 14-26.

Shaman, D., 1996. India's Pollution Regulatory Structure and Background, The World Bank. (http://www.worldbank.org/nipr/india/india-back.htm)

USAEP, 1996. Industry and Environment in Asia: US-Asia Environmental Partnership. (http://www.usaep.org)

Welford, R. and Bhargava, S., 1996. Corporate Strategy and the Environment: The Theory, Richard Welford (Ed.), Corporate Environmental Management: Systems and Strategies, London: Earthscan, 13-32.

Dwivedi, O.P. and Khator, R., 1995. India's Environmental Policy, Programs and Politics, O.P. Dwivedi and D.K. Vajpeyi (Eds.), Environmental Policies in the Third World: A Comparative Analysis, Westport, CT: Greenwood Press, 47-70.

NEERI, 1995. Carrying Capacity Development Planning of NCR Region, MoEF Project.

Porter, M. and Linde, C., 1995. Green and Competitive: Ending the Stalemate, *Harvard Business Review*, 120-34.

Roberts, Michael, 1995. Indian Public Demands Industry Responsibility, Chemical Week, 157(1), 89.

Schoemaker, P. and Schoemaker, J., 1995. Estimating Environmental Liability—Quantifying the Unknown, *California Management Review*, 29-61.

Johannson, L., 1994. Profits from Green Ventures, *The Management Accounting Magazine* (CMA), 68(7), 17-21.

Court ORDER, 1992. India—M.C. Mehta *vs.* Union of India, WP 4677/1985 (1992-05-15) (Delhi Land Use Case: Stone Crushers) 05-15-1992, M.C. Mehta *vs.* Union of India and Others, Writ Petition (Civil) No. 4677 of 1985, (Kuldip Singh, K. Ramaswamy J.J.), 15-05-1992 ORDER.

Stikker, A., 1992. Sustainability and Business Management, *Business Strategy and the Environment*, 1(3), 1-8.

NEERI, 1991. Report on Air Pollution Mitigation System in Lime Kilns, CSIR Thrust Area Project.

Ehrlich, P. and Ehrlich, A., 1990. *The Population Explosion*, London: Arrow Ltd.

Tiwary, J. and Singh, B., 1990. Industrial Development and its Changing Pattern in India during the Plan Period, The Indian Economic Association, 73rd Annual Conference, Muzaffarpur, 46-54.

Rajendran, S., 1989. Institutional Support to SSIs, The Hindu Survey of Indian Industry, M/s. Madras, India: Kasturi & Sons, 279-83.

Nyati, K., 1988. Problems of Pollution and its Control in Small Scale Industries, New Delhi, India: Friedrich Elbert Foundation.

Cornell, B. and Shapiro, A., 1987. Corporate Stakeholders and Corporate Finance, Financial Management, 16, 5-14.

CPCB, 1981. Report on Comprehensive Industry Document for Lime kilns.

Shreve, R.N., 1956. *The Chemical Process Industries*, McGraw Hill Co. Inc., New York, 128139.

Hazardous Air Pollutants

6

N.P. Thacker

Preamble

Toxic air pollutants, also known as hazardous air pollutants are those compounds that are known or suspected to cause cancer and other serious health effects. The health effects include damage to immune system, developmental, respiratory and other health problems. EPA has identified about 188 air pollutants (www.epa.gov/ttn/atw/allabout.html). Examples of toxic air pollutants include persistent organic pollutants (POPs): pesticides, HCB, polychlorinated biphenyls (PCBs), dioxins and furans, benzene, toluene, xylene, ethyl benzene, methylene chloride, vinyl chloride, mercury, etc. People are exposed to toxic air pollutants in different routes including through air, food and drinking water.

Sources

The toxic air pollutants are discharged from natural as well as manmade sources. Natural sources include forest fires, volcanic eruptions while man made sources include various manufacturing processes like chemical and pesticide manufacturing, cement factories, incinerators and furnaces, etc.

Current Status

In India, amongst the various hazardous air pollutants, limited data are available in case of volatile organic compounds (VOCs), polycyclic aromatic hydrocarbons (PAHs) and data are not available on unintentional POPs contamination in air.

Volatile Organic Compounds (VOCs)

The most dominant VOCs are benzene, toluene, ethyl benzene and xylene (BTEX). Further, the hydrocarbon VOCs are grouped under methane and non-methane categories. In Delhi, 52 VOCs in residential areas, 59 in commercial areas and 61 in industrial areas have been identified (Srivastava *et al.,* 2005). Concentration levels of total VOCs in Delhi are presented in Table 1.

Table 1. VOCs in Delhi

Sl. No.	Areas	Concentration (Total VOCs) (µg/m³)
1.	Residential	23.74-162.68
2.	Commercial (Karol Bagh, Lajpat Nagar)	259.7-1369.42
3.	Industrial (Jhilmil, Okhla and Mayapuri)	174.7-656.8

The concentration of the VOCs—benzene, toluene, o-xylene and p-xylene were found to be in the range of 147-3465.3, 42.5-3770, 108.1-2347.8 and 56.6-437.8 µg/m³ respectively in 2000 during winter and summer seasons in the traffic air of Kolkata city (Mukherjee *et al.,* 2003).

Mercury (Hg)

Mercury present in the atmosphere today is due to releases from anthropogenic activities since many years, 85% of mercury emissions come from stacks, primarily power plants and municipal and medical waste incinerators. Natural sources include volcanoes, evaporation from soil and water surfaces, degradation of minerals and forest fires.

The major sources of mercury pollution in India are chlor-alkali industries, industrial processes in thermal power plants, steel and cement industries, coal fired power and heat production, mercury-containing products such as thermometers, blood pressure equipment, pesticides, dental amalgam, and waste incineration processes.

Presently, mercury consumption in the Indian chlor-alkali sector is at least 50 times higher than the global average. This sector alone contributes to about 40 per cent of the total mercury pollution in the country. The average specific mercury loss from a mercury cell plant in India is 142 gms per metric tonne (MT) of caustic soda (NaOH) produced. At this rate, the caustic chlorine sector alone emits about 79 metric tonnes of mercury every year.

Sources of Anthropogenic Releases of Mercury

Releases of Mercury Impurities:

- Coal-fired power and heat production (largest single source to atmospheric emissions).
- Energy production from other fossil carbon fuels.
- Cement production (mercury in lime).
- Mining and other metallurgical activities involving the extraction and processing of virgin and recycled mineral materials, e.g., production of: iron and steel, ferromanganese, zinc, gold and other non-ferrous metals.

Releases from Intentional Extraction and Use of Mercury:

- Mercury mining.
- Small-scale gold and silver mining (amalgamation process).
- chlor-alkali production.
- Use of fluorescent lamps, various instruments and dental amalgam fillings.
- Manufacturing of products containing mercury, for example: thermometers, manometers and other instruments, electrical and electronic switches.

Releases from Waste Treatment, Cremation, etc. (originating from both impurities and intentional uses of mercury):

- Incineration of municipal, medical and hazardous wastes.
- Cremation.

Mercury (Hg) Levels in India

Recent studies have shown that the total mercury pollution potential from coal in India is estimated to be 77.91 tonnes per annum, if average concentration of mercury in coal is assumed to be 0.272 ppm. About 59.29 tonnes of mercury per annum is mobilised from coalfired thermal power plants alone. The five super thermal power plants in the Singrauli area, which supply 10 per cent of India's power, are responsible for 16.85 per cent or 10 tonnes per annum of total mercury pollution from power generation.

Environmental Mercury concentration in Chembur, Mumbai is almost three times higher during the dust storm when compared to normal conditions of 0.93 ± 0.66 mg cm^3. Most of the Hg was present in gaseous form. In certain areas, the average precipitation of Hg was 82 ìg/l. List of Mercury emitters in India are presented in Table 2.

Table 2. Mercury Emitters in India

	Consumers	Tonnes/annum
Mercury in air and flyash of power plant	Coal based thermal power plants	60
Mercury in air and dust	Cement manufacturing plant	NA
Mercury in air	Burning of mineral oil and disposal of medical waste	NA
Mercury in air and effluents	Disposal of municipal solid waste	NA

Polycyclic Aromatic Hydrocarbons (PAHs)

Prevailing concentration of different PAH compounds at different stations in Kolkata, Canning and Moutorh during Diwali are presented in Table 3. (http://www.cpcb.nic.in/Highlights05/ch-4.htm)

Table 3. Average Concentration of PAH Compounds (ng/m^3) at Different Locations in India

PAH Compounds	Canning	Kolkata	Moutorh	Behala	Kumirpukur	JD Park
Napthalene	3.53	1.86	0.91	3.11	1.71	0.76
Phenanthrene	0.28	0.45	BDL	0.55	0.36	0.43
Fluorene	0.93	1.05	2.83	1.30	0.77	1.09
Pyrene	1.65	2.84	5.19	3.85	2.04	2.62
Benz(a) Anthracene	5.88	13.29	20.81	25.87	—	8.36
Chrysene	1.08	4.43	27.57	7.07	—	3.11
Benz(e) Pyrene	2.10	4.39	37.25	6.52	2.02	4.65
Benz(a) Pyrene	8.31	10.84	19.31	16.74	5.40	10.39
DiBenz Anthracene	0.76	1.61	1.73	3.61	0.13	1.10
Benz(ghi) Perylene	2.73	4.23	19.54	6.07	2.03	4.60

Benzo(a) pyrene (BaP) is the most toxic pollutant among PAHs group of compounds and is also carcinogenic. The levels of BaP and PM10 in ten major Indian cities have been reported in Table 4 [NEERI, 2002]. There is a National Air Quality Monitoring (NAQM) Programme, under which NEERI is regularly monitoring BaP along with other air quality parameters for selected cities.

Table 4. Particulate Matter (PM10) and BaP in Metro Cities (Manuel *et al.*, 2004)

Sl. No.	Cities	Seasons	Concentration range BaP (ng/m³)	Concentration range PM 10* (μg/m³)	Regulatory limits (CPCB) PM10 (μg/m³)
1.	Chennai	Winter	8-118		
		Summer	4-96	64-88	(i) Residential – 100
		Monsoon	4-84		
2.	Delhi	Winter	17-233		
		Summer	10-128	168-285	(ii) Industrial – 150
		Monsoon	5-131		
3.	Kolkata	Winter	15-230		
		Summer	3-236	134-247	(iii) Sensitive – 75
		Monsoon	5-243		
4.	Mumbai	Winter	14-187		
		Summer	6-179	90-188	
		Monsoon	11-132		
5.	Nagpur	Winter	17-248		
		Summer	9-124	67-145	
		Monsoon	9-83		

*Annual average

Dioxins and Furans and Polychlorinated Biphenyls (PCBs)

There are 75 dioxins (PCDDs), 135 furans (PCDFs) and 209 PCBs. All differ from one another by the location and number of chlorine atoms attached to the molecule, and their varying degree of toxicity. Only 17 of all 210 dioxins and furans and 10 of 209 PCBs congeners have four lateral chlorine atoms and are of concern. Data on dioxins and furans and PCBs in air in India are not available. However, the facilities for measuring them are now being developed in the country. The work on the measurement of their existing levels in air is in progress. Dioxins and furans quantification is now in progress at NEERI, Nagpur, RRL, Trivandrum and NIOH, Ahmedabad.

Future Challenges

The future challenges and policy measures include:

- Application of eco-friendly methods like avoiding use of chemical pesticides and opting for bio-pesticides.
- Adopting elemental chlorine free bleaching to reduce the formation of dioxins and furans and promoting the use of environmental friendly paints, dyes, etc. in order to reduce VOCs in air.
- Formulation of health based policies.
- Tools and strategies to mitigate mercury pollution that has an impact on human health and environment.
- To formulate appropriate immediate and long-term national action plan as to reduce manmade mercury and PCBs release.

REFERENCES

Srivastava, A., Joseph, A.E., Patil, S., More, A., Dixit, R.C. and Prakash, M., 2005, Air Toxics in Ambient Air of Delhi, Atmospheric Environment, 39(1), 59-71.

Manuel, J.A., Phadke, K.M., Joshi, S.D., Lalwani, A. and Kumar, A., 2004. Levels of Benzo(a) pyrene in Ambient Air of Indian Metropolis, Indian Association for Environmental Management, 31, 107-12.

Mukherjee, A.K., Bhattacharya, S.K., Ahmed, S., Roy, S.K., Roychowdhury, A. and Sen, S., 2003. Exposure of Drivers and Conductors to Noise, Heat, Dust and VOCs in State Transport Special Buses of Kolkata city, Transportation Research Part D; Transport and Environment, 8 (1), 11-19.

SEDAC, Report of the third meeting of the Ozone Research Managers by WMO/UNEP. (http://sedac.ciesin.org/ozone/WMO/india.html)

OSC. (http://osc.edu/research/pcrm/transport/hotspot_daytime.html)

USEPA. Technology Transfer Network, About Air Toxics. (www.epa.gov/ttn/atw/allabout.html)

CPCB, 2005. Highlights on Environmental Research, Air Pollution Source Apportionment Studies. (http://www.cpcb.nic.in/Highlights05/ch-4.html)

Volatile Organic Compounds in Air

7

Anjali Srivastava

Preamble

Volatile Organic Compounds (VOCs) are one of the major group of pollutants of concern. VOCs are part of the large hydrocarbon family, a vast array of aliphatic, aromatic hydrocarbons, their halogenated derivatives, alcohols, ketones and aldehydes. VOCs have the property of getting converted into vapour or gas without any chemical change. They are highly reactive hydrocarbons and participate in atmospheric photochemical reactions. Some of them have negligible photochemical activity; however, they play an important role as heat trapping gases in atmosphere.

In the presence of oxides of nitrogen and sunlight, VOCs form ozone and other products. Oxidation of VOCs by reaction with hydroxyl radicals is the main removal process. The oxidation of complex organic molecules leads to the fragmentation, production of a range of reactive free radicals and more stable smaller molecules such as aldehydes (Atkinson, 2000). VOCs are major cause of concern, firstly due to their role in the formation of ground level ozone and smog and secondly due to some of them being carcinogenic, mutagenic and teratogenic in nature. Adverse effects of ozone on human health, crop viability and yields are well documented. Wide range of VOCs implies wide range of reaction rates, which means large range of transport distances. Many VOCs have low reactivity and thus longer atmospheric lifetime and can be classified as Persistent Organic Pollutants (POPs). Some VOCs are included in Hazardous Air Pollutants (HAPs) by category of their toxicity.

Sources

Many VOCs are of natural origin, while many owe their existence to anthropogenic activities. Natural sources of VOCs include forests, termites, oceans, wetlands, Tundras and volcanoes. Estimated global emission of biogenic VOCs is 1150 Tgyr-1 (Guenther *et al.*, 1995).

The anthropogenic sources of VOCs consist of vehicular emissions, petroleum products, chemicals manufacturing industries, painting operations, varnishes, coating operations, consumer products, petroleum handling, auto refinishing, cold clean degreasing, printing inks, dry-cleaning, etc.

International Scenario

Trans-boundary concerns regarding VOCs arise due to their ability of long range transport, distribution and accumulation in various components of environment, their toxic nature and significant contribution from natural sources.

In 1979, Geneva convention on Long Range Transboundary Air Pollution (LRTAP) was established within the framework of the United Nations Economic Commission for Europe (UNECE) (http://www.apis.ac.uk/overview/regulations). Under this convention in 1991, the protocol on the control of emission of VOCs and their trans-boundary fluxes was adopted and came into force on 29th September, 1997 following ratifications by National Governments. Subsequently, multi effect of air pollutant Gothenburg Protocol was adopted on 30th November, 1999 and signed by 31 countries. This protocol set new targets for emissions cuts by 2010 for SO_2, NOx and VOCs.

India is not a signatory to any one of these protocols. However, India prepared Ecomark Criteria for 14 products specifying VOC contents [http://www.envfor.nic.in/legis/legis.html (EcoMark Schemes)]. Researchers (Srivastava *et al.*, 2000; Mohan Rao *et al.*, 1997, 1988; Khan *et al.*, 2000; Ghoshal *et al.*, 2002; Chattopadhyay *et al.*, 1997; Srivastava *et al.*, June 2000; Pandit *et al.*, 1990, 2001; Pandhy *et al.*, 2000; Maji *et al.*, 2003; Varshney *et al.*, 1998; Srivastava *et al.*, 2004a, 2005a) have monitored VOCs in some urban areas. Source apportionment of VOCs in Mumbai and Delhi has been reported using CMB-8 model (Srivastava *et al.*, 2004b, 2005b). It is reported that evaporative emissions are the major contributors to VOCs in ambient air. Emission inventory of VOCs from anthropogenic sources in India in 1998 is reported as 21 million metric tonnes of VOCs per annum (Varshney, 1998b). At present, NEERI is preparing an emission inventory of VOCs in four metropolitan cities namely Delhi, Mumbai, Kolkata and Chennai based on secondary data. Studies on VOC emissions are very few.

Most Serious Concerns

In India, the most serious concerns are:

- Generating data on ambient VOCs levels.
- Inventory of VOC emissions.
- Need for legislations limiting organic solvents use in certain industries (printing etc.).
- Need for legislations on control of VOC emissions in petrol transportation and handling.
- Research on development of methods for recovery of solvents in industries.
- Meeting stricter and emerging global standards.

Significant emission of VOCs from natural sources necessitates control of VOC emissions from anthropogenic sources. Large number and types of sources warrant specific control/management of VOC emissions. Atmospheric reactivity of VOCs refers to the impacts of individual VOCs on formation of ozone. Regulatory policies, thus need to consider the reactivity of VOCs besides their toxicity.

Future Challenges

Major scientific issues required to be addressed immediately include the following:

- Atmospheric chemistry of VOCs.
- Computational / modelling methodology.
- Emissions inventorization.
- End of life cycle assessment (volatility and fate).
- Effect of environmental conditions on fate and reactivity.
- Reactivity assessment.

In the long run, the persistent VOCs are likely to pose environmental and health hazards by virtue of their accumulation in environmental media. Significant technological challenges lie in meeting existing and emerging international standards of VOC emissions. This requires addressing control of large number of VOCs with different chemical properties and reactivity at source.

REFERENCES

Srivastava, A., Joseph, A.E., Patil, S., More, A., Dixit, R.C. and Prakash, M., 2005a. Air Toxics in Ambient Air of Delhi, *Atmospheric Environment*, 39, 59-71.

Srivastava, A., 2005b. Variability in VOC Concentration in An Urban Area of Delhi, *Environmental Monitoring and Assessment*, Proof available online.

Srivastava, A., Joseph, A.E., More, A. and Patil, S., 2005c. Emission of VOCs at Urban Petrol Retail Distribution Centres in India (Delhi and Mumbai), *Environmental Monitoring and Assessment*, Proof available online.

Srivastava, A., Sengupta, B. and Dutta, S.A., 2005d. Source Apportionment of Ambient VOCs in Delhi City, *Science of the Total Environment*, 343(1-3): 207-20.

Srivastava, A., Joseph, A.E. and Nair, S., 2004a. Ambient Levels of Benzene in Mumbai City, *International Journal of Environmental Health and Research*, 14(3), 215-22.

Srivastava, A., 2004b. Source Apportionment of Ambient VOCs in Mumbai City, *Atmospheric Environment*, 38, 6829-43.

Maji, C.S. and Bhaskar, A.N., 2003. Pollution Preventing Anionic Lithographic Inks, *Journal of Hazard Material*, 105(1-3), 103-19.

Ghoshal, A.K. and Manjare, S.D., 2002. Selection of Appropriate Technique for Recovery of VOCs: An Analysis, *Journal of Loss Prevention in the Process Industries*, 15, 413-21.

Pandit, G.G., Srivastava, P.K. and Mohan Rao, A., 2001. Monitoring of Indoor Volatile Organic Compounds and Polycyclic Aromatic Hydrocarbons Arising from Kerosene Cooking Fuel, *Science of Total Environment*, 279, 159-65.

Srivastava, P.K., Pandit, G.G., Sharma, S. and Mohan Rao, A.M., 2000. Volatile Organic Compounds in Indoor Environments in Mumbai India, *The Science of the Total Environment*, 225, 161-68.

Khan, F.I. and Ghoshal, A.K., 2000. Removal of VOCs from Polluted Air, *Journal of Loss Prevention in the Process Industries*, 13, 527-45.

Srivastava, A.K., Pandit, G.G., Sharma, S. and Mohan Rao, A.M., 2000. Ambient Level of Benzene and Other Aromatic Hydrocarbons in Mumbai, Proceedings of National Symposium on Environment, Bangalore University, 7-10 June 2000.

Pandhy, P.K. and Varshey, C.K., 2000. Total Non Methane Volatile Organic Compounds in the Atmosphere of Delhi, *Atmospheric Environment*, 34, 577-84.

Atkinson, R., 2000. Atmospheric Chemistry of VOCs and NOx, *Atmospheric Environment*, 34, 2063-101.

Mohan Rao, A.M. and Pandit, G.G., 1998. Concentrations of C2-C5 Hydrocarbons in Atmospheric Air at Deonar Bombay in Relation to Possible Sources, *Atmospheric Environment*, 22, 395-401.

Varshney, C.K. and Pandhy, P.K., 1998a. Total Volatile Organic Compound in Urban Environment of Delhi, *Journal of Air Waste Management Association*, 48 (5): 448-53.

Varshey, C.K. and Pandhy, P.K., 1998b. Emissions of Total Volatile Organic Compounds from Anthropogenic Sources in India, *Journal of Industrial Ecology*, 2, 93-105.

Mohan Rao, A.M., Pandit, G.G., Sain, P., Sharma, S., Krishnamoorthy, T.M. and Nambi, K.S.V., 1997. Non Methane Hydrocarbons in Industrial Location in Bombay, *Atmospheric Environment*, 31, 1077-085.

Chattopadhyay, G., Samanata, G., Chatterjee, S. and Chakraborti, D., 1997. Determination of Benzene, Toluene and Xylene in Ambient Air of Calcutta for Three Years During Winter Environmental Technology, 18, 211-18.

Guenther, A., Hewitt, C.N., Erickson, D., Fall, R., Geron, C., Graedel, T., Harley, P., Klinger, L., Lerdau, M., McKay, W.A., Pierce, T., Scholes, R., Steinbrecher, R., Tallamraju, R., Taylor, J. and Zimmerman, P., 1995. A Global Model of Natural Volatile Organic Compound Emissions, *Journal of Geophysical Research*, 100(D5): 8873-91.

Pandit, G.G. and Mohan Rao, A.M., 1990. Evaluation of Auto Exhaust Contribution to Atmospheric C2-C5 Hydrocarbons at Deonar, Bombay, *Atmospheric Environment*, 24A, 811-13.

APIS. (http://www.apis.ac.uk/overview/regulations)

MOEF, Eco-marks Schemes. (http://www.envfor.nic.in/legis/legis.html)

Air Pollution and Health Impacts

8

J.S. Pandey

Preamble

There is an increasing and growing evidence linking air pollution to acute and chronic illnesses. For quantifying impact of air pollution on human health, normally three types of approaches are used: meteorological approach, risk-based (exposure, dose rates) approach and epidemiological approach. The first (meteorological) approach confines itself to the emissions of various air pollutants, their transport, transformation and removal from the environment. The second (risk-based) approach examines the pathways through which air pollution enters the body and estimates the amount of exposure and risk. The third (epidemiological) approach studies health outcomes like hospital admissions, disease rates, mortality and potential mechanisms of biological effects. Although research results have been amply reported in regard to all these three approaches, there has been very little effort in the direction of attempts to integrate the information, which is available on all the three approaches (Middlebrook *et al.*, 2004). Once there is an integration between all the three approaches, it will be useful not only in recent and ongoing research activities connected with air pollution and its control, but also in ultimate health risk (HR) assessment and studies, that would generate public awareness and help in averting and mitigating the overall risks. Policy decisions for reducing pollutant-concentrations, in general, are based on health impact studies.

Current Status

So far as India is concerned, it accounts for nearly one third of the total 150 million asthma afflicted people worldwide. There are reports which suggest linkages between air pollution and tuberculosis (Khalakdinal *et al.*, 2004). India ranks highest in terms of premature deaths due to outdoor and indoor air pollution. Recent findings also suggest that 30% of Delhi's population suffers from respiratory disorders due to air pollution and the incidence of respiratory diseases in the city is as high as 12 times the national average. Common disease symptoms are eye irritation, throat infection, respiratory discomfort, skin ailments, impaired hearing, chest disease, excessive carboxy haemoglobin, and annoyance with noise (http://www.delhitrafficpolice.nic.in/art8.htm). In terms of suspended particulate matter (SPM), Delhi is the fourth most polluted city in the world (http://www.gisdevelopment.net/application/ natural_hazards/overview/nho0019.htm).

Constantly rising population levels, their haphazard distribution and growth and the consequent rise in the levels of infrastructure needed to support them are the main reasons behind this pollution rise. Haphazard industrial development and transportation activities are equally responsible (Srivastava, 2004). Vehicular emissions, in particular, are estimated to account for 60 to 70% of total air pollution (CPCB, 2004). Diesel (generating) sets and vehicles, particularly auto-rickshaws are some of the major sources contributing to air pollution.

Health Impacts of Air Pollution

Air pollutants have long-term effects on respiratory systems and result in decreasing lung-function specially in children (Neuberger *et al.*, 2002). Chronic damage of small airways results in lower endexpiratory flow rates. A long-range research jointly conducted by the National Institute for Environmental Health Sciences and the State of California points out that children belong to the maximum risk category in the urban cities. There are two possibilities through which air pollution may affect lungs (http://mumbai.usconsulate.gov/wwwhwashnews2439.html). Chronic inflammation is one, with air pollutants irritating small airways on daily basis. The other possibility is that the air pollutants might dampen the growth of alveoli (tiny air bags in the lungs).

Immediate Concerns

A committee of experts setting international priorities in Environmental Epidemiology identified asthma incidence as a high priority issue to be investigated (WHO, 1994; Chowgule *et al.*, 1998). Under the circumstances, there is a strong need now for carrying out health risk assessment studies for the cities, which are suffering from serious pollution-related health problems. This involves generation of site-specific data connected with environmental pollution and associated socio-economic profiles. There is also a need for characterizing exposure profiles within every age-group (Wallace, 1987; Pandey *et al.*, 1992, 1993; Pandey *et al.*, 2004). This means generation of more detailed information through intensive monitoring and household surveys.

Emerging Issues

As far as airborne particulate matter (PM) is concerned, it is a mixture of many chemical species (pollutants) in solid and liquid forms. Particulate matter less than 10 microns in diameter is referred to as PM10, and PM less than 2.5 microns is referred to as PM2.5. While trying to understand the role of particle size and composition in terms of its health risks, there are strong evidences supporting the fact that smaller particles are more harmful. Studies have also shown that particles of different sizes vary in their respiratory tract deposition, movement, clearance, and consequent retention time in the human body. Suspended particulate matter (SPM) may also contain fungal spores and pollen, which induce several allergic responses. Because of their acidic nature, sulfate and nitrate ions can even damage respiratory tract (http://www.worldbank.org/sarurbanair).

In urban areas, particulate matter mainly contains metals, organic compounds, materials of biological origin, secondary particulate matter (sometimes even in ionic form) and pure or elemental carbon. The particle core, which often forms the bulk of urban particulate matter mainly comprises elemental carbon.

Many organic compounds can lead to mutations and thus can be cancerous. A major fraction of diesel particles are smaller than 1 micron in diameter. They consist of a carbonaceous core with large surface area on which many organic compounds like carcinogenic polycyclic and nitro-polycyclic hydrocarbons are adsorbed. A recent study by the United States Environmental Protection Study has linked diesel emissions with high cancer risks (USEPA, 2002).

Future Challenges

Since in most of the cities the air is unhealthy, the most important issue is how severely it affects the human-health through the dose inhaled (HEI, 2002). Dose when gets combined with its likely responses, indicates the ultimate health risk (HR). Quantification of health risk (HR) at a given location would require following information (Hall, 1996):

- ambient air quality data for all pollutants of interest.
- exposure-response and/or dose-response functions specific to the local population and pollutant mix.
- demographic data including population, its spatial distribution and age-specificity.
- time-activity profiles (occupancy factors) for different age-groups of population in different regions (residential, industrial, commercial, etc.).

REFERENCES

CPCB, 2004. National Ambient Air Quality Status 2002. Central Pollution Control Board (Ministry of Environment & Forests, India), 2.

Khalakdinal, A., Pokhrel, A., Jenny, A., Kumar, R., Balakrishnan, K., Prasad, R., Verma, S. and Smith, K.R., 2004. South Asian Multi-center Study of Indoor Air Pollution and Tuberculosis: Design and Protocols. BAQ-2004, Agra, India, December 6-8, 2004. (http://www.cleanairnet.org/caiasia/1412/propertyvalue-16289.html)

Middlebrook, A., Turner, J. and Solomon, P.A., 2004. Preface: Special Issue of Atmospheric Environment for Particulate Matter: Atmospheric Sciences, Exposure, and the Fourth Colloquium on PM and Human Health, *Atmospheric Environment* 38, 5179-81.

Pandey, Jai Shanker, Kumar, Rakesh and Devotta, Sukumar, 2004. Health Risk Assessment for Air Pollution Management in India: A Case Study of Delhi City. BAQ-2004, Agra, India, December 6-8, 2004. (http://www.cleanairnet.org/caiasia/1412/propertyvalue-16289.html)

Srivastava, A., 2004. Source Apportionment of Ambient VOCs in Mumbai City, *Atmospheric Environment*, 38, 6829-43.

HEI (Health Effects Institute), 2002. HEI Perspectives: Understanding the Health Effects of Components of the Particulate Matter. Mix: Progress and Next Steps. (http://www.healtheffects.org/Pubs/Perspectives-2.pdf)

Neuberger, M., Moshammer, H. and Kundi, M., 2002. Declining Ambient Air Pollution and Lung Function Improvement in Austrian Children, *Atmospheric Environment*, 36, 1733-36.

USEPA, 2002. Health Assessment Document for Diesel Exhaust, EPA/600/8-90/057F.
(http://cfpub.epa.gov/ncea/cfm/recordisplay.cfm?deid=29060)

Chowgule, R.V., Shetye, V.M., Parmar, J.R., Bhosale, A.M., Khandagale, M.R., Phalnitkar, S.V. and Gupta, P.C., 1998. Prevalence of Respiratory Symptoms, Bronchial Hyperactivity, and Asthma in a Megacity. *American Journal of Respiratory and Critical Care Medicine*,158, 547-54.

Hall, J.V., 1996. Assessing Health Effects of Air Pollution, *Atmospheric Environment*, 30, 743-46.

Pandey, J.S., Pimparkar, S. and Khanna, P., 1993. Health Exposure Assessment and Policy Analysis, *International Journal of Environmental Health Research*, 3, 161-70.

Pandey, J.S., Pimparkar, S. and Khanna, P., 1992. Micro-Environmental Zones and Occupancy Factors in Jharia Coal-Field: PAH-Health Exposure Assessment, *Journal of Environmental Systems*, 21(4): 349-56.

Wallace, L.A., 1987. Total Exposure Assessment Methodology (TEAM) Study, Summary and Analysis, USEPA, Washington, D.C.

WATER

Water Resources

M.V. Nanoti

9

Preamble

India is rich in water resources, being endowed with a network of great rivers and vast alluvial basins to hold groundwater. Conditions, however, vary widely from region to region. There are some chronically drought-affected areas, whereas there are others, which are frequently subject to damage by floods. On the whole, due to the pressure of rapid population growth, the available resources of water are being developed and depleted at a fast rate and the situation seriously underlines the need for taking up integrated plans for water conservation and utilization for every agro-ecological area to meet the increasing demands of irrigation, water harvesting, human and livestock consumption, expanding industry, hydro-electric power generation, recreation, navigation and other uses.

Water Resources

Precipitation in the form of rain and snowfall provide over 4000 km^3 of fresh water to India, most of which returns to the oceans via many large rivers which flow across the subcontinent. A portion of this water is absorbed by the soil and is stored in underground aquifers. A much smaller percentage is stored in inland water bodies, both natural (lakes and ponds) and man-made (tanks and reservoirs).

Surface Water

Keeping in view the variations in the intensity of rainfall, country has been divided into 35 meteorological sub-divisions. The resources potential of the country, which occurs as natural run-off in the rivers is about 1869 km^3 as per the basin wise latest estimates of Central Water Commission (CWC, 2000), considering both surface and groundwater as one system. Ganga-Brahmaputra-Meghna system is the major contributor to total water resources potential of the country. Its share is about 60% in total water resources potential of various rivers. Based on 1991 census, the per capita availability of water works out to 2208 m^3 per annum. Due to various constraints of topography, uneven distribution of resources over space and time, it has been estimated that only about 1122 km^3 of total potential of 1869 km^3 can be put to beneficial use, 690 km^3 being due to surface water resources. Again, about 40% of utilizable surface water resources are presently in Ganga-Brahmaputra-Meghna system. In majority of river basins, present utilization is significantly high and is in the range of 50 to 95% of utilizable surface resources. But in the rivers such as Narmada and Mahanadi percentage utilization is quite low. The corresponding values for these basins are 23% and 34% respectively.

The distribution pattern of water resources potential in the country indicates that as against the national per capita annual availability of water as 2208 m^3, the average availability in Brahmaputra and Barak is as high as 16589 m^3 while it is as low as 360 m^3 in Sabarmati basin. Brahmaputra and Barak basin with 7.3% of geographical area and 4.2% of population of the country has 31% of the annual water resources. Per capita annual availability for rest of the country excluding Brahmaputra and Barak basin works out to be about 1583 m^3. Any situation of availability of less than 1000 m^3 per capita per annum is considered by international agencies as scarcity conditions. Cauvery, Pennar, Sabarmati, East-flowing rivers and West-flowing rivers are some of the basins which fall under this category.

Groundwater

Groundwater represents one of the most important water sources in India and accounts for over 400 km^3 of the annual utilizable resources in the country. Due to the highly variable nature of the climate, groundwater has become a popular alternative for irrigation and domestic water use across India.

Reliance on groundwater resources is particularly strong where dry season surface water levels are low or where wet season flows are too disruptive to be easily tapped. In addition to being accessible, groundwater quality is generally acceptable in most areas and presents a relatively safe source of drinking water for Indians in rural and urban centres.

The presence and availability of groundwater varies greatly with changes in topography, subsurface geology and the prevailing climate in the region. In some areas, groundwater exists in deep aquifers while in others, the water is stored near the surface. The location of the aquifer also affects its recharge rate and its susceptibility to pollution and overuse.

In general, the mountainous and hilly regions in the north and west do not allow adequate infiltration and as a consequence, groundwater is mostly limited to valleys and other low lying areas. In the peninsular part of the country, the underlying geology limits the formation of large continuous aquifers. Groundwater is, therefore, scattered where fissures permit adequate storage or is found in shallow depressions near the surface. As a result, the overall yield potential in this region is low although some areas may have medium to high potential depending on the local hydrogeology. Coastal regions are usually rich in groundwater owing to the largely alluvial terrain, but the aquifers risk being easily contaminated by saltwater ingress due to over pumping. The alluvial tract of the Gangetic plain, which extends over 2000 km across central and northern India has the best potential for groundwater extraction in the country. This large area possesses many favourable characteristics for groundwater storage and recharge, and the yield over most of the region has been estimated as moderate to high.

Surface Water Resources

The most reliable method to assess the surface water resources is through an accurate and continuous study of water flows through rivers and streams over a period of years. The runoff records of the various river systems in India are incomplete and the assessment of the water resources from actual gauged discharge records is not possible for the whole country. However, the rainfall records are available for a long period and attempts have, therefore, been made to roughly assess the water resources on the concept that the natural run-off (assuming that there is neither any upstream development on the surface or groundwater resources nor any increment or decrement in the groundwater storage) is equal to the total volume of precipitation minus the volume of water lost into the atmosphere through evapo-transpiration.

Dr. A.N. Khosla, an eminent engineer, evolved the following relationships for estimating the run-off as a function of rainfall and temperature, assuming that the temperature is the main determinant of evapo-transpiration:

$R_m = P_m - L_m$ ....(1)

$L_m = 0.0481\ T_m$......(2)

where

R_m = monthly run-off, cm

P_m = M_p monthly rainfall, cm

L_m = monthly evaporation losses, cm

T_m = mean monthly temperature, °C.

On the basis of this relationship, he estimated the mean annual natural run-off (normally referred to as the surface water resources) of the six water-resources regions of the country as 167.23 mham. The region-wise distribution is given in Table 1.

Table 1. Region-wise Distribution of Surface Water Resources in India

Name of Basin	Catchment area (mha)	Average annual precipitation (cm)	Total precipitation (mham)	Mean temperature (°C)	Average annual run-off (mham)
Indus Basin	35.40	56	19.82	12.6	7.94
Ganga Basin	97.60	111	108.34	16.8	48.96
Brahmaputra Basin	50.62	122	61.76	8.2	38.08
West-flowing Rivers	49.16	122	59.98	25.5	31.06
East-flowing Rivers	121.03	100	131.92	26.1	41.19
Loni & Ghaggar Basin	16.80	39	4.87	26.2	—
Total	370.61	549	386.69	—	167.23

Later, based on the specific analysis of the stream-flow data from 80 sources, carried out by the Central Water and Power

Commission, the Irrigation Commission (1972) has worked out the water resources of the country at 178 mham. This figure includes, besides surface run-off, groundwater run-off (effluent discharge) which is contributed by the groundwater storage as the base flow of the rivers (as shown later, the present figure of groundwater run-off is roughly assessed at about 45 mham) as well as the sub-surface run-off contributed by rainfall and seepage from surface water resources, but excludes whatever influent recharge (tank storage) that might have taken place due to the flood flows.

The above estimate represents the average annual river flow. All this flow cannot be utilized owing to the highly variable character of the flow and other limitations imposed by the physiographic factors. The concentration of flow in a four-month period necessitates the construction of storage works, if the flood flows passing to the sea are to be intervened and utilized. However, suitable sites for the construction of storage dams are limited. The utilizable quantum also depends on the availability of suitable sites for the diversion of water, the quality of water being suitable, the proximity of water to the land fit for cultivation and the dependability of flow. In India, the irrigation projects are usually designed for 75% dependability, so that the designed quantity of water is available for at least 75% of the time in the year. In view of the limitations imposed by these considerations, the total utilizable flow has been estimated at 66.6 mham, by Irrigation Commission (1972).

Region-wise details of the total available and the utilizable surface water resources are given in Table 2.

Table 2. Region-wise Details of the Total Available and the Utilizable Surface Water Resources

Sl. No.	Name of Basin	Average annual run-off (in mham)	Utilizable flow (in mham)
1.	Indus Basin	7.70	4.93
2.	Ganga Basin	51.00	18.50
3.	Brahmaputra Basin	54.00	1.23
4.	West-flowing Rivers	28.80	6.92
5.	East-flowing Rivers	34.80	33.80
6.	Luni & Ghaggar Basin	1.70	1.22
	Total	178.00	66.60

With further growth of irrigation, it is likely that sub-surface flow contributing to the total annual run-off may appreciably increase, thus augmenting the present availability of surface water resources. This increase, however, may be compensated for by the anticipated reduction in the groundwater run-off owing to the continued development of the groundwater resources. It is, therefore, visualized that the total surface water resources may remain, more or less, steady at the same level as at present.

Groundwater Resources

The assessment of groundwater resources is much more difficult, as it involves the evaluation of the various hydrologic components within the framework of a complex geological environment. There are as yet numerous gaps in the available information and the stage has not reached when any precise basin-wise assessment of the resources can be made except in highly generalised terms.

An attempt was made by the Central Working Group (constituted for working out the Fourth Five-Year Plan proposals for Minor Irrigation in 1968) to work out the rainfall contribution of groundwater recharge by classifying the geological formations in India into 7 categories and assuming, on the basis of the best-informed judgement, different percentage factors for the groundwater recharge for these various categories. The categories and the percentage factors assumed are indicated in Table 3.

Table 3. Assumed Percentage Rainfall Contributing to Groundwater Recharge in Different Rock Formations of India

Sl. No.	Rock formation	Assumed percentage rainfall contributing to groundwater recharge
1.	Hard-rock formation & Deccan traps	10
2.	Consolidated rocks (sandstone)	5-10
3.	River alluvia	15-20
4.	Indo-Gangetic alluviam	20
5.	Coastal alluvia	10-15
6.	Western Rajasthan sand dune	2
7.	Intermontane valleys	15-20

On the above basis, the average contribution to groundwater recharge in the country as a whole works out to 10% of the total rainfall. Since then, some more reliable regional estimates (though still based on shortcut generalised methods) have been made by the state groundwater organisations, and it is felt that the figure of 10% contribution, on the average, may be somewhat conservative. A figure of 12.5% is considered to be more realistic.

Reckoning that the total annual volume of rainfall is about 400 mham, on the basis of mean rainfall of 120 cm over the geographical area of 328 mham, the mean rainfall contribution to the groundwater recharge comes to 50 mham.

As mentioned before, the underground aquifers are supplemented from sources other than rainfall, seepage from canals, water-courses, field channels, ponds, tanks, return flow from irrigation in the fields and the influent recharge from rivers. As assessed later, the total volume of surface water presently utilized for irrigation is about 23.5 mham. It is visualized that about 35% of this total volume diverted for irrigation would pass as deep percolation to groundwater storage. Hence, the addition to the groundwater recharge on this account may be taken as about 8.2 mham. The total volume of groundwater presently used for irrigation is estimated at about 10.5 mham. Since, in the case of groundwater presently used for irrigation, the loss in the long conveyance systems is not involved and the efficiency of irrigation is somewhat better, it may be more realistic to assume a figure of 25% as the deep percolation contributing to groundwater recharge. Addition to groundwater recharge on this account, will thus be 2.75 mham. The total contribution to groundwater recharge from these sources may thus be taken as 10.95 (say 11 mham).

Assuming that about 4 mham, maybe the influent recharge from the flood flows, the total mean availability of groundwater resources at present may be reckoned as about 65 mham. It is roughly visualized to be accounted for as below:

1. Withdrawal use:

(i) Irrigation	10.5 mham
(ii) Other uses	1.5 mham
(iii) Total	12.0 mham

2. Loss due to evapo-transpiration from the water table — 8.0 mham
3. Groundwater run-off (influent recharge to rivers and streams) — 45.0 mham

Total — 65.0 mham

According to the assessment of the National Commission of Agriculture, the contribution from surface water resources to the groundwater recharge is likely to be increased to about 35 mham on the full development of irrigation, 25 mham from seepage from canals, tanks, etc. and 10 mham as influent recharge (to be increased as a result of special measure for induced recharge). Adding the contribution of rainfall, the total ultimate groundwater resources may be of the order of 85 mham. Reckoning that about 50 mham will be lost as groundwater run-off and evapo-transpiration, the ultimate usable groundwater resources may be assessed at 35 mham. The above assessment does not include the additional supplies that may temporarily be obtained by the mining of water in certain regions.

REFERENCE

CWC, 2000. Water and Related Statistics, Central Water Commission, Ministry of Water Resources, Govt. of India, New Delhi.

Water Quality

10

M.V. Nanoti

Preamble

A world water development report of the United Nations has categorized India among the worst countries with respect to poor quality of water and expressed concern on India's ability and commitment to improve the situation. The asian rivers are most polluted in the world with three times as many bacteria from human waste as the global average. These rivers also have 20 times more lead than those of the industrialized countries, states the report.

The report ranks 122 countries according to the quality of the water as well as their ability and commitment to improve the situation. Belgium is considered as the worst basically because of the low quantity and quality of its groundwater due to heavy industrial pollution and poor treatment of wastewater. Morocco, India, Jordan, Sudan, Niger, Burkina Faso, Burundi, Central African Republic and Rwanda follow it.

Attributing this to "inertia at leadership level", the report entitled "Water for People, Water for Life" observes that "the global water crises will reach unprecedented levels in future with growing per capita scarcity of water in many parts of the developing world", it further observes that water resources will steadily decline because of population growth, pollution and expected climate change.

"Globally the challenge lies in raising the political will to implement water related commitments," says the report, "Water professionals need a better understanding of the broader social, economic and political context, while politicians need to be better informed about water resource issues. Otherwise, water will continue to be an area for political rhetoric and lofty promises instead of sorely needed actions."

The list of the countries with best quality is headed by Finland, followed by Canada, New Zealand, United Kingdom, Japan, Norway, Russian Federation, Republic of Korea, Sweden and France.

The top 10 water rich countries are French Guyana, Iceland, Guyana, Surinam, Congo, Papua New Guinea, Gabon, Solomon Islands, Canada and New Zealand. The poorest countries in terms of water availability are Kuwait followed by Gaza Strip, United Arab Emirates, Bahamas, Qatar, Maldives, Libyan Arab Jamahiriya, Saudi Arabia, and Malta Singapore.

The report further states that by the middle of this century, at worst seven billion people in 60 countries will be facing with water scarcity, at best 2 billions in 48 countries, depending on factors such as population growth and policy making. Climate change will account for an estimated 20% of this increase in global water scarcity.

Water Quality Issues

The deterioration of water quality in various water bodies is due to a number of factors, listed in Table 1.

Table 1. Factors Responsible for Deterioration in Water Quality

Rivers	Groundwater	Lakes and reservoirs pollution
• Human activities such as agriculture, deforestation • Release of untreated domestic or industrial wastes • Uncontrolled discharge of industrial wastewater • River eutrophication • Salinization • Changes in river hydrology	• Unsewered domestic waste • Disposal of liquid urban and industrial waste • Disposal of solid domestic and industrial waste • Cultivation with agrochemicals • Salinity from irrigation • Geological formations	• Contaminants, in solution inflow or adsorbed onto particulate matter, from the rivers draining into the lake • Groundwater systems flowing into lakebeds • Wet and dry atmospheric deposition of contaminants on the lake surface

The types and sources of water contamination include "point" sources of pollution which usually refer to wastes being discharged from a pipe, and "non-point" sources, which means all other sources such as storm water run-off (which picks up oils and other contaminants), irrigation (which carries fertilizers and pesticides into groundwater), leaks from storage tanks and leachate from disposal sites. The non-point sources are technically the most difficult to regulate in India. Water pollution takes place from three main sources, viz. domestic sewage, industrial effluents and runoff from activities such as agriculture. Water pollution from domestic and human wastewater causes many severe water borne diseases. The problem of water pollution due to industries is because of the inadequate measures adopted for effluent treatment.

Access to safe drinking water remains an urgent need as only 70.5% of the households in the urban area and 8.7% in rural areas receive organised piped water-supply and the rest have to depend on surface and groundwater which are untreated (Statistical Abstract of India, 1998). The diseases commonly caused due to contaminated water are diarrhoea, trachoma, intestinal worms, hepatitis, etc. The most common contamination in the water is from the disease bearing human wastes, which is usually detected by measuring fecal coliform levels. Inadequate access to safe drinking water sanitation facilities, leads to higher infant mortality and intestinal diseases.

Surface Water Pollution Status

The Central Pollution Control Board has been monitoring water quality of national aquatic resources in collaboration with concerned State Pollution Control Boards at 507 locations (CPCB 1999a), of which 430 stations are under MINARS (Monitoring of Indian National Aquatic Resources), 50 stations are under GEMS (Global Environmental Monitoring Systems) and 27 stations under the YAP (Yamuna Action Plan). The polluted stretches identified in some of the major rivers (Table 2) are based on regular monitoring. The water quality and desired water quality have been classified as A, B, C, D, and E, which reflect the best designated usage of the water. The water quality monitoring results obtained during 1998 indicate that organic and bacterial contamination still continues to be the critical source of pollution in Indian aquatic resources.

Table 2. List of Polluted River Stretches

River	Polluted Stretch	Desired class	Existing class	Critical parameters	Possible source of pollution
Sabarmati	Immediate upstream of Ahmedabad up to Sabarmati Ashram	B	E	DO, BOD, Coliform	Domestic and industrial waste from Ahmedabad
	Sabarmati Ashram to Vautha	D	E	DO, BOD, Coliform	Domestic and industrial waste from Ahmedabad
Subarnarekha	Hatia Dam to Bharagora	C	D/E	DO, BOD, Coliform	Domestic and industrial waste from Ranchi and Jamshedpur
Godavari	Downstream of Nasik and Nanded	C	D/E	BOD	Wastes from sugar industries, distilleries and food processing industries
Krishna	Karad to Sangli	C	D/E	BOD	Wastes from sugar industries and distilleries
Sutlej	Downstream of Ludhiana to Harike	C	D/E	DO, BOD	Industrial wastes from hosieries, tanneries, electro-plating and engineering industries and domestic waste from Ludhiana and Jalandhar
	Downstream of Nangal	C	D/E	Ammonia	Wastes from fertilizer and chloralkali mills from Nangal
Yamuna	Delhi to confluence with Chambal	C	D/E	DO, BOD, Coliform	Domestic and industrial waste Delhi, Mathura and Agra
	In the city limits of Delhi, Mathura and Agra	B	D/E	DO, BOD, Coliform	Domestic and industrial wastes from Delhi, Mathura and Agra
Hindon	Saharanpur to confluence with Yamuna	C	D	DO, BOD, Toxicity	Industrial and domestic wastes from Saharanpur and Ghaziabad

(Contd...)

River	Polluted Stretch	Desired class	Existing class	Critical parameters	Possible source of pollution
Chambal	Downstream of Nagda and downstream of Kota	C	D/E	BOD, DO	Domestic and industrial waste from Nagda and Kota
Damodar	Downstream of Dhanbad	C	D/E	BOD, Toxicity	Industrial wastes from Dhanbad, Durgapur, Asansol, Haldia and Burnpur
Gomti	Lucknow to confluence with Ganges	C	D/E	DO, BOD, Coliform	Industrial wastes from distilleries and domestic wastes from Lucknow
Kali	Downstream of Modinagar to confluence with Ganges	C	D/E	BOD, Coliform	Industrial and domestic wastes from Modinagar

Source: CPCB, 1999 (reported in MoEF, 2001).

Groundwater Pollution Status

Groundwater Pollution is usually traced back to four main sources i.e., industrial, domestic, agricultural and over-exploitation. The last category mainly accounts for seawater intrusion. Studies carried out in India reveal that one of the most important causes of groundwater pollution is unplanned urban development without adequate attention to sewage and waste disposal. Industrialisation without provision of proper treatment and disposal of wastes and effluents is another source of groundwater pollution. Excessive application of fertilizers for agricultural development coupled with over-irrigation and intrusion due to excessive pumping of fresh water in coastal aquifers are also responsible for groundwater pollution. Leachates from compost pits, animal refuse, garbage dumping grounds, nutrient enriched return irrigation flows, seepage from septic tanks, seepage of sewage, etc. has adversely affected the groundwater quality in several parts of India.

Thirteen states in India have been identified as endemic to fluorosis due to naturally occurring fluoride bearing minerals. These are A.P., Gujarat, Haryana, Orissa, Punjab, Rajasthan, Tamil Nadu, U.P., Karnataka, M.P., Maharashtra, Bihar and

Delhi. There are nearly half million people in India suffering from ailments due to excess of fluoride in drinking water. The groundwater in Assam and Orissa has high iron content, resulting in homochromatosis, due to prolonged accumulation of iron in the body. Many states (Rajasthan, Gujarat, Punjab and Haryana) have high incidence of salinity in groundwater. Arsenic contamination in groundwater has been reported in shallow aquifers in many parts of West Bengal.

Arsenic contamination in groundwater resources has been discovered since 1988. The worst scenario is existing in West Bengal where 9 districts out of total 18 districts have been affected by arsenic in the groundwater. The contamination level varies from few ppb to more than 300 ppb in many wells. The affected districts are Murshidabad, Nadia, North 24 Parganas, Malda, where arsenic levels were found in excess of 50 ppb in 24, 17, 21, 9 blocks respectively. In South 24 Parganas, Kolkata, Howrah and Hoogly districts arsenic level in excess of 50 ppb were observed in about 28.8% samples (out of total 8139 samples), 6% samples (out of total 3516 samples), 11.6% (out of total 1280 samples), 15.4% samples (out of total 90 samples) respectively. In Bardhaman district, 40 villages were having arsenic in excess of 50 ppb. In Baliya district of U.P. and Buxar district of Bihar, maximum level of arsenic goes beyond 1000 ppb and 300 ppb respectively. In Rajnandgaon district of Chhattisgarh, 0.88 ppb arsenic was recorded.

Status of Groundwater Pollution in India is presented in the Table 3.

Water Treatment

Treatment of water from various sources, for making it suitable for drinking, industrial, household purposes has been a major challenge. Tube well water was heavily promoted and developed as a safe and environmentally acceptable alternative to microbiologically unsafe untreated surface water. However, in mid-1990s, the Indian government stepped up its efforts to fight back against arsenic, a poisonous element contaminating drinking water in many regions of the country. UNICEF was one of the first international relief agencies, to join the government's efforts, by introducing technologies for groundwater treatment.

Table 3. Status of Groundwater Pollution in India

Pollutant	State	Places of occurrence
Salinity	Maharashtra	Amravati, Akola
(Inland)	Bihar	Begusarai
	Haryana	Karnal
	Rajasthan	Barmer, Jaisalmer, Bharatpur, Jaipur, Nagaur, Jalore & Sirohi
	U.P.	Mathura
Salinity	Andhra Pradesh	Vishakhapatnam
(Coastal)	Orissa	Puri, Cuttack, Balasore
	West Bengal	Haldia & 24 Parganas
	Gujarat	Junagarh, Kachch, Varahi, Banskanta & Surat
Flouride	Kerala	Palaghat.
	Andhra Pradesh	Cuddapah, Guntur, Nalgonda, Nellore, Chittoor, Anantpur
	Gujarat	Banskanta, Kachch & Amreli
	Haryana	Hissar, Kaithal & Gurgaon
	Orissa	Bolangir, Bijapur, Bhubaneshwar and Kalahandi
	Punjab	Amritsar, Bhatinda, Faridkot, Ludhiana & Sangrur
	Rajasthan	Nagaur, Pali, Sirohi, Ajmer & Bikaner
	Tamil Nadu	Chengalput, Madurai
	U.P.	Unnao, Agra, Aligarh, Mathura, Ghaziabad, Meerut & Rai Bareli
Sulphide	Orissa	Balasore, Cuttack & Puri
Iron	U.P.	Mirzapur
	Assam	Darrang, Jorhat, Kamrup
	Orissa	Bhubaneshwar
	Bihar	E. Champaran, Muzaffarpur, Gaya, Monger, Deoghar & Madhubani
	Rajasthan	Bikaner, Alwar, Bharatpur
	Tripura	Dharamnagar, Kailasanar, Ambasa, Amarpur & Agartala
	West Bengal	Madnipur, Howrah, Hoogly and Bankura
Manganese	Orissa	Bhubaneshwar, Athgaon
and	U.P	Muradabad, Basti, Rampur & Unnao
Arsenic	West Bengal	Malda, Murshidabad, Nadia, 24 Parganas
Nitrate	Bihar	Patna, East Champaran, Palamu, Gaya, Nalanda, Nawada and Banka
	Andhra Pradesh	Vishakhapatnam, East Godavari, Krishna, Prakasam, Nellore, Chittoor, Anantpur, Cuddapah, Kurnool, Khammam and Nalgonda

(Contd...)

Pollutant	State	Places of occurrence
	Delhi	Naraina, Shahdara (Blocks)
	Haryana	Ambala, Sonepat, Jind, Gurgaon, Faridabad & Hissar
	Himachal Pradesh	Kulu, Solan, Una
	Karnataka	Bidar, Gulbarga and Bijapur
	Madhya Pradesh	Sihore, Bhopal & West & Central Part of State
	Maharashtra	Jalna, Beed, Nanded, Latur, Osmanabad, Sholapur, Satara, Sangli and Kolhapur
	Punjab	Patiala, Faridkot, Firozpur, Sangrur & Bhatinda
	Rajasthan	Jaipur, Churu, Ganganagar, Bikaner, Jalore, Barmer, Bundi and Sawaimadhopur
Chloride	Tamil Nadu	Coimbatore, Penyar and Salem
	West Bengal	Uttar Dinajpur, Malda, Birbhum, Murshidabad, Nadia, Bankura and Purulia.
	Karnataka	Dharwad, Belgaum
	Madhya Pradesh	Bhind, Shajapur and Sehore
	Maharashtra	Sholapur, Satara, Amravati, Akola & Buldana
	Rajasthan	Barmer, Jaisalmer, Jodhpur & Jalore
	West Bengal	Contai, Digha, Haldia
Zinc	Andhra Pradesh	Hyderabad, Osmania University campus
	Delhi	R.K. Puram
	Rajasthan	Udaipur
Chromium	Punjab	Ludhiana

The precarious balance between growing demands and supplies bring forth the importance of maintaining quality of both surface and groundwater. In the face of very large scale re-use of water, unless the return flows are of reasonably good quality, very serious problems of quality degradation would occur both in surface and groundwater. Correction of quality degradation, particularly, in groundwater is a very difficult process.

Treatment of wastewater is essentially a very costly proposition, while this is inescapable in the future, the costs could be a very serious constraint in encouraging treatment. In particular, treatment of domestic sewage for all the growing urban centres would be somewhat impracticable, considering that most municipalities are not financially self-sustainable, until cheaper alternatives for human wastes disposal could be evolved. Such alternatives are available by way of oxidation ponds, waste stabilization ponds, use of upflow anaerobic sludge blanket (UASB) technology, duck weed pond technology, utilization of raw or partially treated sewerage for forestry, artificial wetlands, roots zone technology, etc.

REFERENCES

Statistical Abstract of India, New Delhi, 1998.

CPCB, 1999a. Water Quality Status and Statistics (1996 and 1997), New Delhi: Central Pollution Control Board, 403.

Groundwater

11

Apurba Gupta

Preamble

Groundwater, one of the most important natural resources of the world, is the major source of drinking water in both urban and rural areas. Agricultural and industrial sectors are largely dependent on the groundwater reserves as well. In India, most of the rivers being non-perennial, this dependable source of water reserve influences the economy to a large extent. However, due to over exploitation, the situation has started changing in the recent years. Being an important and integral part of the hydrological cycle, its availability and distribution depend on the rainfall and recharge conditions and indiscriminate and unplanned withdrawal can cause irreversible environmental and economic problems over a long-term.

Groundwater is accessible to a large number of users. It can provide cheap, convenient, individual supplies. It is generally less capital-intensive to develop, and does not depend upon controversial mega-water projects. Groundwater development is also generally self-financing. It is largely private development and ensures automatic cost recovery. Compared to surface water, groundwater offers better insurance against drought because of the long lag between changes in recharge and responses in groundwater levels and well yields.

Irrigation with groundwater is also generally more productive. Compared to surface water irrigation, groundwater is produced at the point of use, needing little transport. It offers individual farmer irrigation "on demand" that few surface systems can offer and because its use entails significant incremental cost of lift, farmers tend to economize on its use and maximize application efficiency by improved irrigation methods.

Current Status

India's total geographic area is 328.76 mha, out of which 197 mha are covered by arable land. About 57 mha are irrigated that constitute 30% of the total cultivable area. Physiographically, the country has a variety of widespread features including mountains, foothills, plains and deserts. Climatically, the variety is abundant as well. The average rainfall, i.e. total precipitation divided by the total land area, is about 1215 mm. However, there is considerable spatial variation in rainfall which ranges from less than 100 mm in the western Rajasthan to more than 2500 mm in North-Eastern India. The rainfall is distributed unevenly not only in space but also in time. Rainfall and associated recharge of groundwater takes place during the four monsoon months of June to September. Within these four months also, most of the rainfall comes in a few spells of intense rain. It is estimated that in Himalayan rivers, where there is some flow due to snowmelt, about 80% of the total annual flow takes place within these four monsoon months. In peninsular rivers, where there is no contribution from snowmelt, monsoon flow accounts for more than 90% of the annual flow.

Due to this uneven distribution of rain in time and space, it is possible to utilize only a small part of it. It is estimated that out of 4000 billion cubic meters (bcm) of precipitation, the annual flow in the rivers is only 1869 bcm out of which only 690 bcm (17%) is currently put to use. The remaining water either undergoes evapotranspiration or groundwater recharge. The basin-wise average flow and groundwater potential i.e. the amount that can be used in a year on a sustainable basis has been estimated and is given in the Table 1.

The annually 'replenishable' groundwater resources of the country have been assessed as 432 bcm (11% of total precipitation). After keeping a provision of 71 bcm/y for domestic and industrial use, 361 bcm/y is available for irrigation. Thus, total utilizable quantity of water is 690 + 432 = 1122 bcm/y (28% of total precipitation). The country also has enormous

Table 1. Basin-wise Average Flow and Groundwater Potential

River Basin	Origin	Length km	Catchment Area Sq.km	Avg. Flow bcm/y	Groudwater Potential bcm
Brahmai with Baitarni	Ranchi (Bihar)	799	39033	28.48	4.05
Brahmaputra	Kailash Range (Tibet)	916+	194413+	585.6	26.55
Cambay Composite	—	—	—	—	7.19
Cauvery	Coorg (Karnataka)	800	81155	21.36	12.3
Ganga	Gangotri (Uttar Kashi)	2525+	861452+	525.02	170.99
Godavari	Nasik (Maharashtra)	1465	312812	110.54	40.65
Indus	Mansarovar (Tibet)	1114+	321289+	73.31	26.49
Krishna	Mahabaleshwar (Maharashtra)	1401	258948	78.12	26.41
Kutch and Saurashtra Composite	—	—	—	—	11.23
Mahanadi	Nazri Town (Madhya Pradesh)	851	141589	66.88	16.46
Mahi	Dhar (Madhya Pradesh)	583	34842	11.02	
Barak-Meghna			41723+		8.52
Narmada	Amarkantak (Madhya Pradesh)	1312	98796	45.64	10.83
Northeast Composite	—	—	—	—	18.84
Pennar	Kolar (Karnataka)	597	55213	6.32	4.93
Sabarmati	Aravalli Hills (Rajasthan)	371	21674	3.81	—
Subarnarekha	—	—	—	12.37	1.82
Tamil Nadu	—	—	—	—	18.22
Tapi	Betul (Madhya Pradesh)	724	65145	14.88	8.27
Western Ghats					17.69
West Flowing Rivers from Tadri to Kanyakumari				113.53	
West Flowing Rivers from Tapi to Tadri				87.41	
West Flowing Rivers of Kutch, Sabarmati including Luni				15.1	
Minor River Basins Draining into Bangladesh and Myanmar				31	
East Flowing Rivers Between Mahanadi and Pennar				22.52	
East Flowing Rivers Between Pennar and Kanyakumari				16.46	
			TOTAL	**1869.37**	**431.44**

Source: Indian Water Resource Society, Delhi : http://www.iwrsdelhi.org/indiaswr.htm

groundwater resources below the lowest level of groundwater fluctuations, known as 'in-storage' groundwater resource which is up to the depth of 450 metres in alluvium and 100 metres in hard rock terrain. This difficult to access resource is estimated as about 10812 bcm.

Overexploitation of Groundwater

A large percentage of the country's GDP is closely tied to the availability of groundwater. Agriculture is contributing to roughly 29% of India's GDP and production from irrigated land is estimated to utilise over 83% of groundwater. Due to rapid and indiscriminate increase in groundwater utilization for irrigation of both food and cash crops, the water levels in several parts of the country have been lowering during the past two decades. India's rapidly rising population and changing lifestyles have also increased the domestic need for water. The picture is same for industrial utilization as well. While at the national level, only 30% of the actual groundwater potential has been harnessed, in a few states such as Punjab, against a critical level of 80%, the level of exploitation is over 98%. Haryana is a close second at 80%, and Tamil Nadu is reaching criticality at 60%. It is evident from the figures above that the groundwater crisis prevailing in several parts of India is not the result of natural factors but caused by human actions.

Groundwater Quality Issues

The quality of groundwater is getting severely affected because of the widespread pollution of surface water. Besides, discharge of untreated wastewater through bores and leachate from unscientific disposal of solid wastes also contaminates groundwater, thereby reducing the quality of fresh water resources.

The quality of groundwater is also getting affected due to overexploitation.

Fluoride contamination: Many states have been identified as endemic to fluorosis due to abundance in naturally occurring fluoride-bearing minerals. These are Andhra Pradesh, Assam, Gujarat, Haryana, Orissa, Punjab, Rajasthan, Tamil Nadu, Uttar Pradesh, Karnataka, Madhya Pradesh, Maharashtra, Bihar, J&K,

Kerala, West Bengal, Meghalaya and Delhi. Nearly half million people in India are suffering from ailments due to excess of fluoride in drinking water.

Iron contamination: In some districts of Madhya Pradesh, Assam and Orissa, groundwater has high iron content.

Water salinity: About 31% of the total area of Rajasthan comes under saline groundwater. Groundwater is saline in the entire Bhakra Canal in Punjab and the lift canal system in south-western Haryana.

Arsenic contamination: High levels of arsenic in groundwater have been reported in the shallow aquifers in some districts of Madhya Pradesh and West Bengal.

Mercury contamination: Certain places in Haryana, Gujarat, and Andhra Pradesh were found to have dangerously high levels of mercury.

Government Initiatives for Management

India has a history of groundwater use since 3000 BC. In the modern times, the first large scale development of groundwater for irrigation was undertaken in 1934 with the installation of 1500 deep tube wells in the Ganga Basin. The latest figures show that there are hundreds of groundwater development projects initiated in canal command and rain-fed areas. The Central Ground Water Board (CGWB) was established in 1972 to formulate the development and management strategies at national level.

The National Water Policy (1987) delineated the strategy for control on the exploitation of groundwater through regulation and an integrated and coordinated development of surface- and ground-water.

More than 15,500 Network Stations of CGWB and more than 30,000 observation wells by the State Groundwater Organizations are in use for periodic monitoring of groundwater level. In 1995, Government of India launched a World Bank aided National Hydrology Project to create a Hydrologic Information System. Currently, it is being implemented in eight states of India. The Project aims at improving the existing data collection network through construction of observation wells, setting up and upgradation of chemical laboratories, establishment of national and state data banks, integration and transmission of data. Under this project so far, 2200 sites spread over eight states have been identified for installation of piezometers. About 2134 piezometers were installed till January 2000. With the strengthening of manpower, infrastructure and better understanding of processes, it is expected that assessment of groundwater potential would be more realistic in future.

Groundwater is considered as a state subject and only State Governments are empowered to enact the law to control and regulate the groundwater exploitation as per the Constitution of India. Realizing the problem of overdraft and mining of aquifers and groundwater pollution, a model bill for groundwater regulation was circulated by the CGWB in the early 1970s. The Bill envisaged the setting up of Ground Water Authority and introduction of a system of licensing for the purpose of extraction and use of groundwater in notified areas, registration of existing users in such notified areas and imposition of penalty for contravening certain provisions. However, barring few states, lot more needs to be done.

International Concerns

World's aggregate groundwater resources appear abundant. Groundwater—both stock and flow constitutes over two-thirds of the world's fresh water resources, even if the glaciers and permanent snow cover (Shiklomanov, 1993; Dutt, 1987) are excluded. If atleast 8 per cent of the 33,000 km^3 (Postel, 1999) annual floodwater that runs off to the oceans recharge the groundwater, it provides a renewable supply of over 2,500 km^3 of groundwater annually. This seems several times more than the world uses today.

In South Asian countries, barring few isolated patches, use of groundwater is substantially below the availability. This is primarily because of high marginal cost of groundwater use in comparison to the water available from canal. Some of the poorest irrigators in South Asia, who purchase pump irrigation from well-owners, commonly pay US 10-14 cents/m^3 of water compared to a fraction of a cent paid by canal irrigators. Thus, in developing countries of Asia and Africa, groundwater development has become the central element of livelihood creation programs for the poor.

Even in US, groundwater is considered as one of the Nation's most important natural resources. It is the source of about 40% of the water used for public supply and provides drinking water for more than 97% of the rural population who do not have access to public water supply systems. Around 30% to 40% of the water used for the multi-billion-dollar agricultural industry in the west comes from groundwater.

Pollution of groundwater resources has become a major concern world over. The pollution of air, water, and land has an affect on the pollution and contamination of groundwater. The solid, liquid, and the gaseous wastes that are generated, if not treated properly, result in pollution of the environment; this affects groundwater too due to the hydraulic connectivity in the hydrological cycle. Water extraction without proper recharge and leaching of pollutants from pesticides and fertilizers into the aquifers has polluted groundwater supplies. In addition, leachates from agriculture, industrial waste, and the municipal solid waste have also polluted surface and groundwater. Around 45 million people the world over are affected by water pollution marked by excess fluoride, arsenic, iron, or the ingress of salt water.

Most Serious Concerns

The country has begun to consider rainwater harvesting and groundwater recharge seriously at all levels. These are at the heart of its massive Integrated Watershed Development Programme, which provides public resources to local communities to treat watershed catchment areas and to construct rainwater harvesting and recharge structures. Trends during the 1990s also suggest a progressive shift of budgetary allocation from irrigation development to water harvesting and recharge.

Across India, some 6.2 mha of rain-fed lands are currently under treatment through 5,200 micro-watersheds, at a whopping cost of Rs. 8 billions (for the year 2001-02). Of the total cultivable area of 142 mha, 89 mha of non-irrigated land requires similar investments. This segment of cultivated area alone contributes a significant 45% of the total foodgrain harvest in the country. With irrigated areas having reached a production plateau of 110 million tonnes, any increase in foodgrain production, from the current level of 200 million tonnes, in the coming decades will have to come from rain-fed agriculture.

Adequate supply of water to the farmers with productive land and potable drinking water to the people without deteriorating the groundwater reserve is the main concern in the country. Available scientific tools and proper investment can lead to achieve the objective since the resource is not exhausted yet.

Artificial Recharge of Groundwater

Artificial recharge of groundwater reservoir has been recognized as one of the important strategies of groundwater management to counter the overexploitation of groundwater. Artificial groundwater recharge reduces or even reverses the declining levels of groundwater, protects fresh groundwater in coastal aquifers against the saline water intrusion and stores surplus surface water including monsoon run-off and wastewater for future use. Artificial groundwater recharge is accomplished mainly through rainwater harvesting, recharging the aquifer with run-off water and vegetative treatments of the catchment. Recharging of aquifers with water is a costly affair. In general, vegetative treatment of watershed helps in increasing the groundwater recharge. Since long distance transport of water for recharge is difficult and expensive, emphasis is being given on *in situ* rainwater harvesting and recharge.

Roof water harvesting for groundwater recharge has given new hopes to overcome the groundwater problems particularly in major Indian cities. Central Ground Water Authority of India is advising the states and municipal bodies to undertake roof top rainwater harvesting and its recharge to groundwater by making it mandatory for every dwelling unit by amending city bye-laws.

Emerging Issues

Development, use and management of groundwater is a complex issue. Knowledge of occurrence of groundwater and its distribution; various considerations like use of static and dynamic components, well interference; developed strategy in relation to stream flow, aquifer yield, overdraft and water quality and knowledge of suitability of various types of groundwater withdrawal structures are very important. Selection of suitable size of pump in relation to the demand is another important aspect of groundwater development and management. Various groundwater management alternatives need to be implemented depending on their suitability.

The groundwater availability in a formation is largely governed by the state of cementation and compaction of the formation. The prevalent rock formation ranging in age from Archaean to recent, controls the occurrence and movement of groundwater. The geological formations encountered in the country may be broadly divided into three categories, namely, the unconsolidated, the semi-consolidated and the consolidated, and the knowledge of their presence can determine the recharge and storage characteristics.

Groundwater resources available for development can broadly be categorised as dynamic and static components. The dynamic component represents the long-term average annual recharge under conditions of maximum groundwater use. Under these conditions, the entire basin area forms the recharge area and groundwater discharge occurs only through the pumping wells. The dynamic groundwater resource is the amount of groundwater available in zone of water level fluctuations. The replenishable groundwater resource is essentially a dynamic resource which is replenished annually or periodically by precipitation, irrigation, return flow, canal seepage, tank seepage, influent seepage etc. Static groundwater resource is the groundwater contained within the permanently saturated zone of the groundwater reservoir, and represents total groundwater reserve minus the dynamic component. This is also defined as the amount of groundwater available in the permeable portion of the aquifer, below the zone of water level fluctuations. Generally, the static component of groundwater is much more than the dynamic component and forms an assured source during drought periods when there is acute shortage of water.

While on the one hand, research in agriculture is evolving irrigation techniques requiring lesser water, hydrogeological investigations coupled with remote sensing data and GIS analyses are now able to quantify the groundwater availability and recharge potential for any area, on the other. Large number of state and national remote sensing centres and private consultants are engaged in such studies all over the country.

Future Challenges

Study of groundwater occurrence, it's distribution and existing level of groundwater development are essential for proper development and utilization of this precious resource. In India, groundwater development and use is restricted to the shallow zone within 50 metres depth and mostly financed through institutional sources and private efforts. The development of deeper zone (50 to 300 metres below ground level) is usually in the public sector for community irrigation. Shallow groundwater structure includes dug wells, dug-cum-bore wells, shallow tube wells, filter points etc. The deeper structure includes heavy duty tube wells and bore wells. Besides the technical expertise required for the proper development, public-private participation is believed to be the essential for its sustenance. Community awareness and participation has proved to be very important for the success of groundwater development.

New Technologies and Approaches

Exploration of groundwater with new tools has resulted in more success stories at optimum investment. They include:

- Mapping the whole country employing remote sensing satellite imageries for identification of water bodies, land use, land cover classes and waterlogged areas. The Indian Remote Sensing (IRS) Linear Imaging and Self-Scanning Sensor (LISS III) providing False Colour Composite (FCC) images have already proved to be very effective for the purpose. Many other imageries in different spectral bands and resolution are also available from National Remote Sensing Agency (NRSA), Hyderabad for the purpose.

- Mathematical modelling of groundwater flow and contaminant transport has become the most effective tool for water resource quantification and aquifer remediation. These can assess land subsidence, seawater Intrusion, water levels in the high plains aquifer, untapped groundwater, studies related to toxic substances in groundwater, crude-oil spill, gasoline spill, low-level radioactive waste and similar other features for groundwater development and protection.
- Application of Geographical Information System (GIS) in water resource and groundwater management is also growing in the recent past. Using the meteorological, geological, geomorphological, landuse and topography data, GIS can suggest optimal location, type and size of the water harvesting structures.
- Techniques to identify and date recently recharged water that are useful in characterizing the susceptibility of hydrogeologic environments to contamination.
- Widely used geochemical models that can be used to improve estimates of the direction and rate of groundwater flow and to predict the transport and fate of inorganic contaminants.
- Geophysical methods for determining the hydraulic properties of fractured rocks. Predicting the movement of water and contaminants through fractured rocks has been one of the most challenging problems in hydrogeology.

Conjunctive use of Surface and Groundwater

Among the major problems with groundwater, excessive application of surface irrigation water in combination with poor sub-surface drainage has created waterlogging and salinity problem in certain canal commands. Conjunctive use of surface and groundwater and subsurface drainage is important for efficient utilization and management of groundwater resource. If planned and designed properly, conjunctive use can enhance the availability of water supplies and ensure advance irrigation in season prior to availability of surface water and provide life saving irrigation when surface water is not available. Conjunctive use of groundwater and surface water in areas having saline groundwater needs careful planning so that the salt water intrusion can be prevented. Planning of conjunctive use project requires proper understanding of the nature and extent of groundwater, estimation of amount of water to be pumped from an area through groundwater balance studies and economic analysis.

Demand Regulation

By adopting suitable cropping patterns in critical and sub-critical areas, the demand of water could be regulated providing more opportunity to the groundwater recharge so that the water table decline is reversed and brought to the earlier level of groundwater reserve. Precision agriculture and water saving crop production technologies should be promoted in such areas. The Agriculture Department may identify suitable cropping patterns for hot-spot areas.

Control of Groundwater Pollution

Non-availability of comprehensive data sets in India is major problem in detection and prevention of groundwater pollution. Central Pollution Control Board and Central Ground Water Board of India are engaged in monitoring of water quality in the country but still the extent and distribution of pollutants across the country has not been quantified. In most of the cases, groundwater monitoring is done in the areas from where the water quality problem has already been reported. Localized groundwater maps for different coverages should be developed and made public. Other than legislation and checks to conserve and improve the quality of groundwater, society itself plays a very important role. During the last decade there has been a rising awareness among the common people on the need for conservation and development of groundwater. Water use has to be integrated effectively with water harvesting using several well-known traditional technologies.

Regeneration of Groundwater Reserves

Renovation of forest tanks in drought-prone regions will have a significant impact on wildlife and forest cover. Similarly, in some urban cities there is a need to regenerate groundwater aquifers because of the high degree of dependence on them for drinking water. Rainwater harvesting schemes have been taken up in many cities and even made compulsory in some of them. Temple tanks need to be renovated and urban wetlands protected. All these will contribute to a rise in the groundwater level and a reduction of salt water ingress. Community awareness and management of freshwater resources should be enhanced. The government should implement effective groundwater legislation and regulations through self-regulation by communities and local institutions. External support agencies should support freshwater resource management. Environmental restoration should be promoted along with household water security.

Improvement in Irrigation Practices

The agriculture sector in India uses 85% of the country's available fresh water. However, irrigation efficiency is only 20-50%. In other words, Indian agriculture wastes up to half of the country's fresh water supply. Although from a basin perspective, much of the wasted water is reused, significant amount of water is wasted primarily due to irrigation inefficiencies. There are inefficiencies on the energy front as well. Agriculture accounts for about 27% of the total electricity consumption in India. Water withdrawal is an energy intensive operation performed throughout the agricultural sector and results in a third of the power consumption in the country being used for the roughly 50% of the national irrigation consumption extracted from groundwater resources. Highly subsidized power supply policies for agriculture have major implications on the overall condition of the power sector and associated water resources. This in turn has affected the power situation of the country in providing unlimited irrigation opportunities.

Conclusion

India is blessed with enough rainwater and water from ice melt and yet the country is passing through water crisis. The effective solution to the freshwater crisis is to integrate groundwater conservation and development activities. The scope extends from water extraction to water management starting at the local level by making communities aware and involved in the activities. Many tools for systems planning are now available. All these must ultimately pave the way for combining conservation of the environment and yet fulfilling the basic water needs of people.

REFERENCES

RIOB, 2002. Groundwater Management in South Asia. (http://www.riob.org/ag2002/GroundWaterAsia.htm)

GOI, 2000. Water and Related Statistics, Central Water Commission, Ministry of Water Resources, Govt. of India, New Delhi.

Chaddha, D.K., 2000. Groundwater Recharge: Option for Groundwater Management, *Journal of Transport and Infrastructure*, 7(4), 44-57.

CGWB, 2000. Groundwater Resource of India, Annual Report, Central Ground Water Board, Ministry of Water Resources, Govt. of India, Faridabad.

CGWB, 1995. Annual Report, Central Ground Water Board, Ministry of Water Resources, Govt. of India, New Delhi.

CPCB, Groundwater. (http://www.cpcb.nic.in/groundwater/ch170703.htm)

Padmanaban, S., Govindarajalu, C. Economic Sector Work on Power Supply to Agriculture: Sub-Task Report on Cost Minimization Through Integrated Agricultural DSM.

USAID. (http://www.usaid.gov/in/UsaidInIndia/Articles28.htm)

Groundwater Contamination—Arsenic

12

V.A. Joshi

Preamble

Arsenic is naturally occurring metalloid ranking twentieth in abundance in earth's crust, fourteenth in seawater and twelfth in human body. Natural factors facilitating introduction of arsenic into surface water are related to geomorphology, tectonic activities and chemical components of water bearing formation (Tong, 2002; Htay, 2004; Fengthong, 2002). Arsenopyrite (FeAsS) is the most common arsenic ore. Arsenate As(V), the oxidized form of arsenic is common in soils that are situated above the water table. Arsenite As (III) is generally found in the soil and sediment below the water table where groundwater is devoid of oxygen and arsenic is in reduced state. Most of the arsenic contamination in West Bengal is found below the water table under reduced condition.

Arsenic is widely distributed throughout the earth's crust. It is introduced in water through dissolution of minerals and ores. Inorganic arsenic occurs in the environment in several forms but in natural water, it is mostly found in trivalent arsenite (AsIII) or pentavalent arsenate (AsV).

Long-term exposure to arsenic causes cancer of the skin, lungs, urinary bladder and kidney, as well as other skin changes such as pigmentation and thickening (hyperkeratosis). Exposure to arsenic through drinking water causes severe disease of blood vessels known as 'black foot diseases' which has been reported in China (Province of Taiwan). This disease has not been observed in other parts of the world (www.Chelationtherapyonline.com/articles/pgg.html). Arsenic in drinking water resulted in

2,00,000-2,70,000 deaths from cancer in Bangladesh alone (NRC, 1998; Smith *et al.*, 2000). The long-term ingestion of arsenic causes cancer of liver, lung, kidney and bladder leading to high mortality (Williams *et al.*, 1998).

Current Status

Arsenic in groundwater has been found to be above the maximum permissible limit (0.05 mg/l) in six districts of West Bengal which covers an area of 34,000 km^2 with population around 30 millions (Chatterjee *et al.*, 1995; Das *et al.*, 1995). The groundwater samples collected from an area of 270 km^2 in Chennai also showed arsenic level above permissible limit (Ramesh *et al.*, 1995).

International Status

There are many other countries like Argentina, Australia, Chile, China, Hungary, Mexico, Peru, Thailand and USA, affected due to arsenic contamination in drinking water. The countries where adverse health effects have been documented are Bangladesh, China, India (West Bengal) and United States of America. As per the studies of British Geological Survey in 1998 on shallow tube wells in 61 of the 64 districts, 46% of samples recorded a concentration above 0.01 mg/l and 27% were above 0.050 mg/l (BGS, 2000). EPA of United States of America has estimated some 13 million population of USA in Western state are exposed to arsenic in water at 0.01 mg/l (USEPA, 2001).

Most Serious Concerns

The most serious concern in water supply is availability of required quantity of safe drinking water to the rapidly growing population. The global fresh water consumption has risen six-fold between 1900 and 1995 which is more than twice the rate of population growth. In addition to this, the problem of biological and chemical contamination is becoming more serious. The chemical contamination due to excess nitrate, fluoride and arsenic is of major concern. The areas in lower Ganges delta region particularly the Bangladesh and six districts of West Bengal (India) are seriously affected due excess arsenic contamination. In West Bengal, it affects an area of 3400 km^2 with a population of

30 millions. More than 800000 people from 312 villages/wards are drinking arsenic contaminated water and amongst them at least 175000 show arsenical skin lesions. The analysis of tube well water, hair, nails, scales, urine and liver tissues show elevated concentration of arsenic. The source of this excess arsenic is geological.

Various social problems are associated with arsenical skin lesions in these districts. Malnutrition, poor socio-economic conditions, illiteracy, food habits have aggravated the arsenic toxicity. If alternate water resources are not utilized, a good percentage of 30 million people may suffer from arsenic toxicity in near future (SOS, arsenic.net).

In Bangladesh, Samta village is different from other villages because nobody wants to marry any girl nor choose bridegroom from the Samta village due to severe arsenic problem. Water of 97% tube wells is unsuitable for drinking due to arsenic contamination. There are hundreds or thousands of such villages in Bangladesh that are not yet known but confronting the worst scenario like Samta village (SOS, arsenic.net).

There are three broad areas where challenges must be effectively dealt to overcome the ongoing arsenic crisis. Efforts need to be made to understand the magnitude of the problem and its causes. The existing technologies for treatment of arsenic contaminated water need to be refined to the point that they are reliable, cost effective and low tech to become locally acceptable under widely varying conditions of the region. It is important to understand the fate of naturally and anthropogenically induced arsenic pollution in the environment and quantify to assess the impact of this emerging crisis.

So far, the research has been carried out on presence of arsenic in groundwater and exposure to arsenic through drinking and cooking. Nowadays, there is widespread use of groundwater for irrigation. The arsenic withdrawn with irrigation water can build up in soil and gets transported into irrigation crops. Phycotoxicity due to increased arsenic in soil and water and its long-term impact on agriculture is another major concern that requires further investigations.

Emerging Issues

Arsenic in groundwater above permissible limit of 0.05 mg/l has been found in six districts of West Bengal. At present, 37 administrative blocks by the side of the river Ganga and adjoining area are affected. The affected area is located in the upper delta plains and mostly in the abandoned meander belt. More than 800000 people from 312 villages/wards are drinking arsenic contaminated water and among them, at least 175000 show arsenal skin lesions. The source of arsenic is geological. Borehole sediment analysis shows high arsenic in only few soil layers and is found to be associated with iron pyrite.

Various social problems arise due to arsenical skin lesions in these districts. Malnutrition, poor socio-economic conditions, illiteracy, food habits and intake of arsenic contaminated water for many years have aggravated the arsenic toxicity. As major water demand in all these districts are met from groundwater, the geochemical reaction, caused by high withdrawal may be the cause of arsenic leaching from source. If alternate water source is not provided, a significant population from 30 millions people of these districts may suffer from arsenic toxicity in the near future.

Future Challenges

The main challenge in arsenic affected area is how to provide the millions of people at risk, chemically safe and aesthetically acceptable alternative source of drinking water. The factors contributing towards the challenge are:

- Identification of arsenic free tube wells in the known affected areas.
- Identification of arsenicosis patients.
- Treatment of the patients with arsenicosis.
- Treatment of arsenic contaminated water where no other safe source exists.
- Identification of arsenic free aquifers, extent of the potential alternative sources, namely, surface water and rain water.
- Habit, culture and customs of the people.

- Economic condition of the people, and their willingness and ability to pay for the service.
- Community motivation to face the challenge collectively.
- Technologies for removal of arsenic.
- Cost.
- Monitoring.
- Research and Development.
- Institutional support.

It has been observed that tube wells exist in any cluster or community. Generally all tube wells in an area are affected by arsenic. Therefore, to find out the unaffected ones in the affected areas and routine monitoring of source to prevent the use of currently affected tube wells after detection of arsenic is difficult.

The single largest contamination due to arsenic has been observed in lower Ganges delta region. The people of this vast region are continuously exposed to arsenic toxicity causing serious health hazards. The arsenic poisoning from the contamination of groundwater is very chronic in nature. Most of the times the victims take casual approach and do not complain about the symptoms until the manifestations are detected through screening. The present experience to identify the arsenic cases are by external manifestation such as melanosis (blackening of skin) and keratosis (hardening of palm and soles) with the history of consuming arsenic contaminated source water. Sometimes persons even after consuming contaminated water does not show any manifestation, but still the chances of having cancer cannot be ruled out.

It is very difficult to differentiate the arsenic manifestation in stages. It is reported that some stages of arsenicosis are reversible, if the person affected is provided with arsenic free water. The use of arsenic free water may probably hinder the deterioration of the symptoms but information on complete recovery from effects is not yet known. Necessary supportive treatment with nutrition improvement may play some role in minimizing symptoms and may help in reversing the melanosis to some extent.

REFERENCES

Htay, L., 2004. Mitigation of Arsenic Contamination in Drinking Water Sources of Myanmar, Training Workshop on the Development of Management and Decision Making Tools for the Mitigation of Contamination of Soils, Crops and Water in the Greater Mekong Subregion, Bangkok, Thailand, 16-18 June 2004.

Fengthong, T., 2002. Lao PDR's Country Report on Development of Effective Management and Decision Making Tools for the Mitigation of Contamination of Soils, Crops and Water in the Greater Mekong Subregion, Training Workshop on the Development of Management and Decision Making Tools for the Mitigation of Contamination of Soils, Crops and Water in the Greater Mekong Subregion, 16-18 June 2004, Bangkok, Thailand, 2004.

Tong, N.T., 2002. Arsenic Pollution in Ground Water in the Red River Delta. ESCAP-IWMI Seminar on Environmental and Public Health Risks due to Contamination of Soils, Crops, Surface and Groundwater from Urban, Industrial and Natural Sources in South East Asia, Hanoi, Vietnam, 10-12 December 2002.

USEPA, 2001. Arsenic in Drinking Water, p. 3, 2001. (www.Chelationtherapyonline.com/articles/p99.htm)

BGS, 2000. Arsenic in Drinking Water. (www.Chelationtherapyonline.com/articles/p99.htm)

WHO, 2000. Arsenic in Your Body Drinking Water for Instances, 2000. (www.Chelationtherapyonline.com/articles/p99.htm)

NRC, 1998. Arsenic in Your Body Drinking Water for Instances. (www.Chelationtherapyonline.com/articles/p99.htm)

Williams, T.M., Rawlins, B.G., Smith, B. and Breward, N., 1998. In-vitro Determination of Arsenic Bioavailability in Contaminated Soil and Mineral Beneficiation Waste from Rob Phibun, Southern Thiland: A basis for Improved Human Risk Assessment, Environmental Geochemistry and Health, 20, 169-77.

Chatterjee, A., Das, D., Mandal, B.K., Choudhury, T.R., Samanta, G.J. and Chakraborti, D., 1995. Arsenic in Groundwater in Six Districts of West Bengal, India: The Biggest Arsenic Calamity in the World, Part 1—Arsenic Species in Drinking Water and Urine of the Affected People. Analyst, 120(3), 643-50.

Das, D., Chatterjee, A., Mandal, B.K., Samanta, G., Chakraborti, D. and Chanda, B., 1995. Arsenic in Groundwater in Six Districts of West Bengal, India; the Biggest Arsenic Calamity in the World Part 2—Arsenic Concentration in Drinking Water, Hair, Nails, Urine, Skin-scale and Liver Tissue (biopsy) of the Affected People. Analyst, 120(3): 917-24.

Ramesh, R., Shivkumar, K., Eswaramoorthi, S. and Purvja, G.R., 1995. Mitigation and Contamination of Major and Trace Elements in Ground Water of Madras City, India, Environmental Geology, 25, 126-36.

Groundwater Contamination—Fluoride

13

V.A. Joshi

Preamble

Fluorine in the combined state constitutes 0.078% of the earth's crust. Fluorine is estimated as the thirteenth in abundance amongst the elements. The fluoride in groundwater is due to its leaching from minerals present in earth's crust.

If the concentration of fluoride in water exceeds 1.5 mg/l, mottling of teeth; called dental fluorosis may occur. Excessive intake of fluoride can result in skeletal fluorosis and other skeletal abnormalities which have been observed normally with consumption of water containing above 3.0 mg/l fluoride (Jackson and Weidmann, 1959). Intake of water with fluoride in excess of 10 mg/l over a long period may result in crippling fluorosis. Drinking water is a major source of fluoride intake which depends upon concentration and daily water intake (McClure, 1943).

Current Status

Fluoride in groundwater is low in Deccan Basalt Province (Western-Central India), highest in the arid and semi-arid belts of Western India and moderately high in parts of semi-arid tracts of the Southern states of India (Rao, 1974).

In India, the problem of endemic fluorosis was first reported in Andhra Pradesh, Tamil Nadu and Uttar Pradesh with 50-100% affected districts. The Punjab, Madhya Pradesh, Maharashtra, Karnataka, Bihar and Haryana have lower intensity of disease with 30-50% of the affected districts (Rajan *et al.*, 1988). An estimated 62 million people in India from 18 states are affected by dental, skeletal and/or non-skeletal fluorosis. The concentration of excess fluoride reported varies from 1-48 mg/l (Susheela, 1999).

International Status

The problem of fluorosis is reported in 23 countries of the world which include USA, Italy, Holland, Spain, France, Germany, Switzerland, China, Japan and several African and South American countries, Africa, Argentina, Australia, Brazil, China, England, Equador, Hungary, Italy, Japan, Korea, Malaya, Manchuria, Mexico, New Zealand, Russia and USA (Susheela, 1999).

Defluoridation Technologies

Since it is widely known that excess fluoride in drinking water causes mottling of teeth in children, many methods have been investigated at national as well as international level for removing excessive fluorides (Maier, 1947, 1953; Savinelli and Black, 1958; Bulusu *et al.*, 1979; Iyengar and Venkobachar, 1997). The Nalgonda technique and activated alumina are field tested technologies. Electrochemical method is a method of generating aluminium hydroxide without chemical addition. The electro-coagulation and electro-dialysis are promising in water treatment because they permit a continuous treatment process and are economical as well as easily automated (Granovskii *et al.*, 1976; Kul'skii, 1986).

Most Serious Concerns

The issues emerging due to rapid growth of population and development of modern technologies is the availability of required quantity of drinking water without any chemical contamination. The global fresh water consumption has risen six folds between 1900 and 1995 which is more than twice the rate of population growth. The countries, which are in high water stress, are Africa, West Asia, India and many other developing countries. In addition to this, problem of biological and chemical contamination is becoming more serious. In India, 62 million people from 18 states are in threat of excess fluoride problem (Susheela, 1999).

A total of 527 plants based on Nalgonda technique and activated alumina for defluoridation of drinking water were installed under Rajiv Gandhi National Drinking Water Mission (RGNDWM) in various states. Out of these, only 77 plants are working. These technologies were critically evaluated by Harcharan Committee constituted by Ministry of Rural Development, Govt. of India which indicated that Nalgonda Technique is a proven technology (Singh, 1995).

An appraisal of the activated alumina technology for defluoridation by Harcharan Committee indicated the necessity for preconditioning of water with acid for lowering the alkalinity and restoration of treated water to neutralise acidity. The need for further research and development was suggested for simplification of process. In ion exchange resin technology, resins used for conditioning and defluoridation of drinking water should be of food grade characterised by low levels of organic leachables during use. Fluoride exchange capacity should be in the range of 0.99-1.5 kg Fluoride/m^3. After treatment, the product water pH should be in the acceptable range as specified by the Bureau of Indian Standards.

As per the committee report, the main reasons for non-functioning of these plants are lack of trained personnel to operate the plant, water quality monitoring to regulate dosages, supervision and guidance by the water supply agencies and funds for operation and maintenance of the plant.

Emerging Issues

With the increasing use of groundwater, the problem of excess fluoride is coming into picture in various states and new states are emerging as fluoride affected states. A fluoride concentration of 18 mg/l, 20 mg/l and 15 mg/l has been reported in Ladakh (J&K), Tattapani (Chhattisgarh) and Bakreshwar (WB) respectively. The fluoride concentration of 8.2-13.2 mg/l has been detected in Kandra (Orissa). In Maharashtra and Himachal Pradesh, the areas identified are Unkesher and Manikaran with a fluoride content of 2.7-6.0 mg/l and 0.2-6.5 mg/l respectively (Sharma, 2003). In view of the above, there is a need for broad concept in development and implementation of a strategy for fluoride and fluorosis control. The water supply sector will have to adopt an innovative, possibly more flexible approach with regard to the interpretation of current standards, norms and design criteria for the planning and execution of water supply schemes. The classification and prioritization of problem areas along with the provision of dual supply systems need to be tried and practiced in the excess fluoride affected areas.

In areas with undulating physiography and remote villages, the highly sophisticated and expensive interventions such as surface water intakes with treatment, regional schemes, defluoridation plants, etc. as per set norms may not be practicable. The type of technology to be adopted for defluoridation may be appropriate and site specific depending on the extent of fluoride contamination, water quality with respect to treatment technology, volume of water to be treated and the awareness and cooperation of the people.

Inadequacy in Management and Control

The problems associated with functioning of Defluoridation plants are inadequacy of management and control and shortage of power supply in rural area. The main problems associated with operation and maintenance of Defluoridation plants are as follows:

- The Operation and Maintenance personnel are transferred or shifted from their jobs after 2-3 years which enable them to get full acquaintance and experience in Defluoridation unit/plant. There should be sufficient overlapping period for handing over during transfer to get acquaintance with Defluoridation plant by transferee.
- Necessary spare parts and accessories are not available at site and breakdown and repairs are not attended to immediately. Log book is not maintained at treatment plant site.
- Funds are not allotted for O&M and are not made available for maintenance of plant from other sources.
- Installations of Defluoridation unit are (Fill and Draw) done without ascertaining the suitability of the raw water quality for defluoridation and proper site selection.
- There is a need for reappraisal of the design criteria and materials for construction of plants to prevent

leakages, corrosion, fissures, peeling of inner linings of the tanks, etc.

- Properly designed dosing system is needed for addition of alum or lime. The stirrers in the reactors should have protective lining to prevent corrosion.
- The functioning of plants is not assessed by water supply department. The Public Health Engineers of the Water Supply Departments, preferably at the junior levels entrusted with the responsibility of the defluoridation plants, should be imparted training on water quality monitoring, defluoridation techniques, and the health hazards due to ingestion of excess fluoride.

Future Challenges

Certain constraints need to be understood regarding the supply of water to fluoride affected areas. All the fluoride affected areas may not have access to rivers in reachable distances. Even if alternate source is available, the implementation of scheme may need certain period and population may have to suffer. In such cases, there is no alternative other than installing defluoridation units—community or domestic. Similarly, the fluoride problem has also been identified not only in rural areas but also in urban areas. Under such conditions, there is a need for further development of technology on Defluoridation at community as well as domestic level. The R&D on defluoridation technologies should continue and field trials of all developed technologies may be conducted until a reliable way of solving the problem is achieved.

The control measures for fluorosis by appropriately designed health survey and water quality assessment in endemic areas and/or habitations where the manifestation of fluoride toxicity is indicated need to be taken up on priority. Once fluorosis and its relation with excessive fluoride in drinking water is established, the control is to be achieved through supply of fluoride free drinking water.

REFERENCES

Susheela, A.K., 1999. Fluorosis Management Programme in India, Current Science, 77(10): 1250-55.

Iyengar, L. and Venkobachar, C., 1997. Defluoridation of Water Using Activated Alumina, Proceeding of National Workshop on Defluoridation Technologies and Fluorosis Control, S.K. University, Anantpur.

Evaluation of the Activities of the Sub-mission, 1995. Control of Fluorosis and Fluorosis Control Cell, Harcharan Singh Committee Report, Ministry of Rural Areas and Employment, Government of India.

Rajan, B.P., Gnanasundaram, N. and Shanthini, R., 1988. Fluoride in Toothpaste: Cause for Concern, Fluoride, 21(4), 167-70.

Kul'skii, I.A. and Strokach, P.P., 1986. Technology of Treatment of Natural Water, Vischa Shk. Kiev.

Bulusu, K.R., Sundaresan, B.B., Pathak, B.N., Nawlakhe, W.G., Kulkarni, D.N. and Thergaonkar, V.P., 1979. Fluorides in Water, Defluoridation Methods and Their Limitations, *Journal of Institution of Engineers* (India), Environmental Engineering Division, 60(1), 1.

Granovskii, M.G., Lavrov, I.S. and Smirnov, O.V., 1976. Electrical Treatment of Fluids Khimiya, Leningrad.

Raghava Rao, K.V., 1974. Incidence of Fluoride in Groundwater, Proceedings of the Symposium on Fluorosis, Hyderabad, 163-73.

Jackson, D. and Weidmann, S.M., 1959. Fluorine in Human Bone Related to Age and Water Supply of Different Regions, *Journal of Pathology Bacteriology*, 76, 451-59.

Savinelli, E.A. and Black, A.P., 1958. Defluoridation of Water with Activated Alumina, *American Journal of Water Works Association*, 50, 33.

Maier, F.J., 1953. Defluoridation of Municipal Water Supplies, *American Journal of Water Works Association*, 45, 879.

Maier, F.J., 1947. Methods of Removing Fluoride from Water, *American Journal of Public Health*, 37, 1559.

McClure, F.J., 1943. *American Journal of Diseases of Children*, 66, 362-69.

Bioremediation of Contaminated Groundwater

14

M.P. Patil

Preamble

The groundwater plays an important role as a decentralized source of drinking water for millions of rural and urban population in India. It is reported (Dinesh Kumar and Shah) that groundwater accounts for nearly 80% of the rural domestic water needs, and 50 per cent of the urban water needs in India. Besides this, groundwater is also a major source of irrigation water. However, the uncontrolled disposal of industrial and domestic wastes (solid wastes, hazardous wastes, industrial effluents, sewage), use of chemical fertilizers and pesticides for agriculture and overexploitation of groundwater have led to the widespread contamination of groundwater and pose significant threats to these resources. Though groundwater is less vulnerable to pollution as compared to surface water resources, the remediation of groundwater is extremely difficult and cost prohibitive, once it occurs. A number of cases of groundwater contamination have been reported in India (Das *et al.*, 1994; Pawar *et al.*, 1998; Datta *et al.*, 1997; Datta *et al.*, 1999; Dinesh Kumar and Shah (http://www.indianngos.com/issue/water/govt/waterqualityissues.htm, http://www.rainwater harvesting.org/Crisis/Groundwater-pollution.htm, http://edugreen.teri.res.in/explore/water/ground.htm), Times of India, Mumbai, 20/05/2002). As reported in these studies, groundwater at many places in India contain fluoride, salinity, heavy metals, pesticides, organic and inorganic substances in concentrations exceeding the Indian Drinking Water Standards (IS – 10500: 1995). Besides these, there are many sites in the country viz. Bichhri in Rajasthan, Ranipet in Tamil Nadu,

Kodaikanal in Tamil Nadu where groundwater and soil are contaminated with toxic substances. Thus, in view of the rapid depletion and widespread contamination of groundwater resources, remediation of contaminated groundwater resources has assumed significant importance in India.

Current Status

Both *in situ* as well as ex-situ options are available for remediation of contaminated groundwater. These options include natural attenuation, bioremediation, phytoremediation, physical barriers, air sparging, air stripping, chemical oxidation, absorption, adsorption, ion exchange, precipitation, coagulation, and flocculation. Among all these options bioremediation is the most cost effective, least harmful and widely applied technology for remediation of variety of contaminants. The bioremediation technologies can be applied to a variety of pollutants viz. organics, heavy metals and inorganic substances, wherein the pollutants are either consumed by the microorganisms (biodegradation) or transformed to non-toxic forms (biotransformation). A number of studies (Grindstaff, 1998; Cunnigham *et al.*, 2001; Knapp and Faison, 1997; Gunther *et al.*, 1994) have demonstrated the successful application of this technology for remediation of groundwater contaminated with various pollutants.

International Concerns

With the advent of recombinant DNA technology in the 1970s, the use of microorganisms that were genetically engineered to use contaminants as a food source was hailed as the future of bioremediation. Indeed, the potential of recombinant DNA technology in bioremediation was recognized when the first patent for a genetically modified organism was issued in 1971 for a Pseudomonas aeruginosa strain that degrades hydrocarbons. However, the deployment of this and other engineered bugs "in the field" is being debated due to uncertainties, legal wranglings and a polarized public opinion concerning their use. Thus, application of genetically engineered and custom designed microorganisms is a major challenge for the international scientific community.

Though the bioremediation technologies have been successfully applied at a number of contaminated sites in developed countries, it is yet to be implemented on a pilot or full scale in India. The major reasons for this include poor enforcement by the regulatory issues, non-availability of skilled manpower and lack of technical and financial resources. Thus, there is an urgent need to address these issues.

Challenges

Though bioremediation is a widely practiced technology for contaminated groundwater, it faces several challenges that need to be addressed in near future. The first one is the complexity of the subsurface environment. Contamination may be confined to one area, or may have fanned out throughout the groundwater in variety of patterns. Thus, delineation of contamination plume and prediction of spread of contamination through detailed hydrogeological assessment will play an important role in successful application of bioremediation technology.

A second major challenge in bioremediation is the selection of the right microorganisms to do the job. Till last decade, identification of microorganisms was carried out by tests that rely on the growth of microorganisms. However, many microorganisms in nature are very hard to culture. Over the past decade, however, scientists have been increasingly successful at using DNA-based identification tests to identify bacteria. These genetic tests are indicating that the growth-based methods have detected only about 1% of all the bacteria that are present in nature, including underground. A clear picture of the underground microbial world is emerging. In the not too distant future, refinements in the genetic identification of bacteria will enable scientists to genetically probe a contaminated site and catalogue the bacterial populations. The dominant populations will be good candidates for bioremediation, having acquired the ability to dine on the toxic compounds present.

Keeping in view the widespread contamination of groundwater resources in the country and the versatility of bioremediation in handling both organic as well as inorganic pollutants, there is a tremendous potential for application of bioremediation technology for restoration of contaminated groundwater resources in the country.

REFERENCES

Cunningham, J., Rahme, H., Hopkins, G., Lebron, C., Reinhard, M., 2001. Enhanced *In Situ* Bioremediation of BTEX-Contaminated Groundwater by Combined Injection of Nitrate and Sulfate, Environmental Science & Technology, 35, 1663-70.

Datta, P.S., Tyagi, S.K., Bhattacharya, S.K., Mookerjee, P., Gupta, N. and Bhatnagar, P.D., 1999. Groundwater NO_3 and F Contamination Processes in Pushkar Valley, Rajasthan as Reflected from ^{18}O Isotopic Signature and ^{3}H Recharge Studies, *Journal of Environmental Monitoring and Assessment*, 56(2): 209-19.

Grindstaff, M., 1998. Bioremediation of Chlorinated Solvent Contaminated Groundwater, U.S. EPA, National Network of Environmental Management Studies.

Datta, P.S., Deb, D.L. and Tyagi, S.K., 1997. Assessment of Groundwater Contamination from Fertilizers in Delhi area Based on ^{18}O, NO_3^- and K^+ Composition, *Journal of Contaminant Hydrology*, 27(3-4), 249-62.

Knapp, R.B. and Faison, B.D., 1997. A Bioengineering System for *In situ* Bioremediation of Contaminated Groundwater, *Journal of Industrial Microbiology Biotechnology*, 18(2-3), 189-97.

Das, D., Chatterjee, A., Samanta, G., Mandal, B., Chowdhury, T.R., Samanta, G., Chowdhury, P., Chanda, C., Basu, G. and Lodh, D., 1994. Arsenic Contamination in Groundwater in Six Districts of West Bengal, India: The Biggest Arsenic Calamity in the World, Analyst, 119(12), 168-70.

Gunther, K., Schlosser, D., Schmauder, H.P. and Rausch, U., 1994. Bioremediation of Contaminated Groundwater, Applied Biochemistry & Biotechnology, 48(1), 11-14.

Dinesh Kumar, M. and Shah, T. Groundwater Pollution and Contamination in India: The Emerging Challenge. (http://www.iwmi.cgiar.org/iwmi-tata/files/pdf/groundpollute4_FULL_pdf)

Indian Standards/Specification for Drinking Water Quality (IS: 10500-1995).

TERI, Water. (http://www.edugreen.teri.res.in/explore/water/ground.htm)

Indian NGO. (http://www.indianngos.com/issue/water/govt/water qualityissues.htm)

CSE, Rainwater Harvesting. (http://www.rainwaterharvesting.org/Crisis/Groundwater-pollution.htm)

Arsenic Crisis

15

Sadhana Rayalu

Preamble

Considering the ubiquitous nature of arsenic in the environment, human exposure to arsenic is inevitable. Exposure can occur via all three principal routes, that is, through inhalation of air, ingestion of food and water, and via dermal absorption. Worldwide, the degree of non-occupational exposure to arsenic varies greatly, being dependent on local geochemistry and the level and proximity of anthropogenic activity. The most important remedial action is prevention of further exposure by providing safe drinking water. In most of the areas, source substitution may be impossible due to non-availability of alternate sources. Arsenic removal may be more appropriate in these situations. The cost and difficulty of reducing arsenic in drinking water increases as the targeted concentration lowers. It varies with the arsenic concentration in the source water, the chemical matrix of the water including interfering solutes, availability of alternative sources of low arsenic water, mitigation technologies, amount of water to be treated, etc. So far dearsenification has been attempted using several techniques including coagulation and filtration (Cheng *et al.*, 1998), ion exchange (Ghurye *et al.*, 1999), membrane methods (Mallevialle *et al.*, 1996). Besides these, possibility of using sorbents like activated alumina (Stumm *et al.*, 1996), rice husk (Lee *et al.*, 1999), iron-oxide coated sand (Joshi *et al.*, 1996), activated carbon (Daus *et al.*, 2004), silica containing iron oxide (Zeng *et al.*, 2003), zeolites (Shevade *et al.*, 2004; Xu *et al.*, 2002) and natural solids (Gonzalez *et al.*, 2001) have been explored globally. Some studies have reported preliminary successes in using packets of chemicals for household treatment. These packets contain mixture of chemicals which remove arsenic as well as act as disinfectant. These packets developed by WHO/PAHO Pan American Centre for Sanitary Engineers and Environmental Sciences in Lima, Peru (CEPIS) has proven successful in Latin America (CEPIS Report, 1968-1998). Some of these researches are promising, but these technologies are still under development and have to be evaluated for field applications.

Current Status

In India, millions of persons in seven districts of West Bengal are drinking groundwater with arsenic concentrations far above acceptable levels (Mandal *et al.*, 1996; Chakraborti *et al.*, 2003). Thousands of people have already been diagnosed with poisoning symptoms, even though much of at-risk population has not yet been assessed for arsenic-related health problems (Chakraborti *et al.*, 2003). Meanwhile, new search across India's Ganges Basin suggest that the crisis in the sub-continent could extend much further than previously thought. According to Chakraborti *et al.* (2004), the Bengal Delta "may be only the tip of the iceberg". Untold numbers of the region's 450 million residents could be exposed to the dangerous levels of arsenic in drinking water. This prompts urgent region-wide well-water analysis. The arsenic problem seems to have intensified during a period of long neglect and this mistake needs to be overcome. Tipped off to a spate of cancer deaths and skin lesions in the village of Semria Ojha Patti in the Indian state of Bihar, an extensive survey of sampled wells in this village has been conducted (Chakraborti *et al.*, 2003). The survey indicated that one in five wells exceeded arsenic levels 30 times more than the stipulated safe level. This maybe attributed to the fact that Bihar is 500 kilometres west of the Bengal Delta and geologically akin to much of the Ganges basin.

There are several organizations in the country that are engaged in different R & D projects related to arsenic. Significant amount of work has been carried out on arsenic removal in the laboratory as well as in the field by the All India Institute of Hygiene and Public Health (AIIHPH), Kolkata and Environmental

Engineering Laboratory, Civil Engineering Department, Bengal Engineering College (Deemed University), Howrah. Besides, the other organisations involved in arsenic removal studies include NML, Jamshedpur, Kalyani University, Center for Study of Man and Environment (CSME), Kolkata and the School of Environmental Studies (SoES), Jadavpur University, Kolkata. Extensive research has been undertaken at NEERI (Pande *et al.*, 1978, 1979, 1980, 1997, 2001; Deshpande *et al.*, 2005) for the development of novel methods for detection and removal of arsenic from drinking water, in addition to extensive survey in arsenic affected areas (NEERI, 2000). R&D pertaining to development of novel hybrid nanomaterials (NHNMs) based on molecular sieves for dearsenification and molecular imprinting agent for arsenic detection is being pursued at NEERI.

Efforts have been made to develop the materials as follows:

- Surface modified zeolites by surface treatment of zeolites using surfactant and quartenary ammonium compounds to develop anionic characteristics.
- Iron treated zeolite by surface treatment of zeolite with iron salts to form iron-hydroxide and metal nanoparticle impregnated material.
- Functionalised treated mesoporous materials by treating MCM-41 with arsenic selective ligands like dimercaptosuccinic acid and dimercaptopropanol.

It is expected that these new materials would provide efficient arsenic removal processes to attain the stringent stipulated standard of 10 ppb in different countries.

International Concerns

The research in Bihar has sparked fears that similar arsenic contamination in Vietnam, Thailand and Taiwan could also be more widespread (Chakraborti *et al.*, 2003). Also many communities rely on groundwater in the United States and are affected by high levels of arsenic in some drinking water supplies. Because the small water systems typically rely on wells for drinking water, while the larger systems typically rely on surface-water sources, arsenic tends to occur in higher levels more often in water used by small communities. The average level measured in US groundwater samples is around 1 ppb, but higher levels are not uncommon (Sharpe *et al.*, 2003).

The environmental health disaster due to excessive arsenic in drinking water is unfolding in many other countries including Argentina, Australia, Bangladesh, Chile, China, Hungary, Mexico and Peru. Reliable data on exposure and health effects are rarely available, but it is clear that there are many countries in the world where arsenic in drinking water has been detected at concentrations greater than the Guideline Value, 0.01 mg/l or the prevailing national standard. Countries where adverse health effects have been documented include Bangladesh, China, India (West Bengal), and the United States of America. According to a British Geological Survey study in 1998 on shallow tube-wells in 61 of the 64 districts in Bangladesh, 46% of the samples were above 0.010 mg/l and 27% were above 0.050 mg/l. When combined with the estimated 1999 population, it was estimated that the number of people exposed to arsenic concentrations above 0.05 mg/l is 28-35 millions and the number of those exposed to more than 0.01 mg/l is 46-57 millions (BGS, 2000). Environmental Protection Agency has estimated that some 13 millions of the population of USA, mostly in the western states, are exposed to arsenic in drinking water at 0.01 mg/l, although concentrations appear to be typically much lower than those encountered in areas such as Bangladesh and West Bengal. Extensive research is under progress worldwide to combat arsenic crisis, particularly to develop promising dearsenification technologies.

The issue of arsenic in drinking water is now recognised as a global problem. Relevant research is being undertaken on many fronts. Technologies designed for small-scale community treatment systems may not be suitable for developing countries, as they are costly and require technical expertise. International donors and aid agencies are funding research into appropriate treatment technologies and techniques that could be developed quickly and effectively in the southern Asia and other effective regions. WHO, for instance, has sponsored a technique called STAR (Stevens Technology for Arsenic Removal) as an effective and inexpensive method for filtering out arsenic from household drinking water suppliers (WHO, 2002).

Most Serious Concerns

The two important issues that need to be addressed immediately include the health effects and bottlenecks or problems associated with existing remediation technologies.

Health Effects

Chronic arsenic poisoning, as occurs after long-term exposure through drinking water, is very different to acute poisoning, typically, includes vomiting, oesophageal and abdominal pain and diarrhoea. Long-term exposure causes cancer of the skin, lungs, bladder, and kidney, as well as other skin changes such as pigmentation and thickening (hyperkeratosis). Increased risks of lung and bladder cancer and arsenic-associated skin lesions have been observed at drinking water arsenic concentrations of less than 0.05 mg/l. Absorption of arsenic through the skin is minimal and thus hand-washing, bathing, laundry, etc. with water containing arsenic does not pose a human health risk (Sharpe *et al.*, 2003). Finally, research continues to probe the health effects of prolonged low-level arsenic exposure. Recently reported work has made important breakthroughs in this area, suggesting mechanisms for direct DNA interactions and co-carcinogenic effects. Other research into biomarkers of exposure seems to confirm that for some level of exposure, some people run a higher risk of developing cancer than others. Such findings should help in developing a better informed public health response to the arsenic issue in both developed and developing countries (NIEHS, 2003).

Limitations of Existing Remediation Technologies

In the backdrop of the widespread public concern due to consumption of groundwater containing high concentration of arsenic, the urgency for delineating appropriate treatment strategy comprising technological innovations for arsenic removal is realized, to overcome the grave threat of chronic arsenic poisoning. Further, it needs to be emphasized that only a few de-arsenification technologies in vogue have been successfully demonstrated in field. Most of the existing technologies have severe limitations including high cost, sludge disposal, chemical handling requirement, waste brine disposal, water rejection in

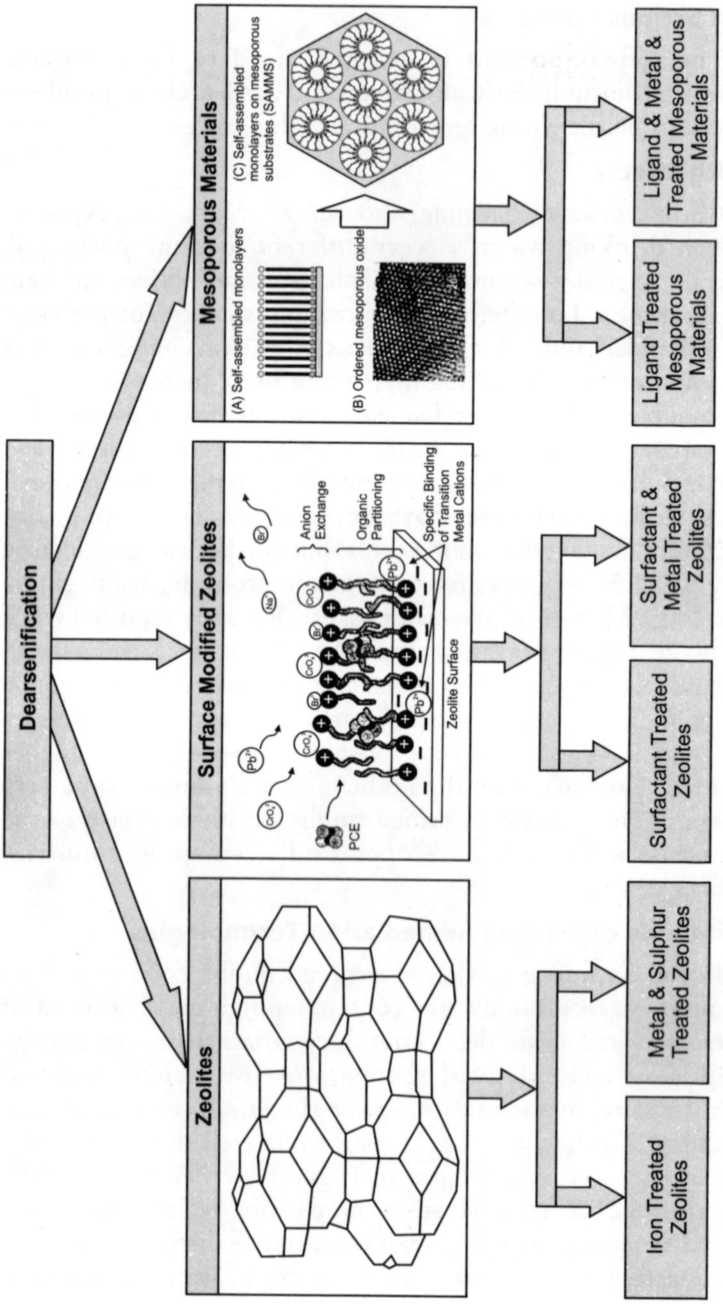
Dearsenification
Zeolites
Surface Modified Zeolites
Mesoporous Materials
(A) Self-assembled monolayers
(B) Ordered mesoporous oxide
(C) Self-assembled monolayers on mesoporous substrates (SAMMS)
PCE
Anion Exchange
Organic Partitioning
Specific Binding of Transition Metal Cations
Zeolite Surface
Iron Treated Zeolites
Metal & Sulphur Treated Zeolites
Surfactant Treated Zeolites
Surfactant & Metal Treated Zeolites
Ligand Treated Mesoporous Materials
Ligand & Metal & Treated Mesoporous Materials

water scarce region, etc. Keeping in view these limitations associated with various technologies like coagulations, adsorption using activated alumina, ion exchange, reverse osmosis, nanofiltration and electro dialysis, it becomes imperative to develop efficient dearsenification technologies based on new processes or advanced functionalised materials.

In this connection, it needs to be reiterated that global interest in improved methods of arsenic removal has led to a rich period of experimentation, in which novel sorbents have been assayed, and existing technologies for arsenic removal modified and/or combined to yield substantially different systems. These novel systems have typically been evaluated only by a handful of researchers, under a narrow range of environmental conditions, for short periods of time. In many cases, advances in removal of arsenic, particularly of arsenite, has been demonstrated, but mechanisms are still poorly understood. Much more research needs to be conducted to identify those technologies that can effectively remove arsenic under real-world conditions, and consistently do so for extended periods of time. The biggest challenges ahead lie with development of:

- Robust affordable household filters for arsenic removal at domestic levels.
- Feasible treatment technologies for removal of arsenic from wells at community level.
- Cost effective and efficient point-of-use and point-of-entry treatment units for arsenic.

The newly developed dearsenification technologies should be able to provide safe drinking water to poor, rural settings at community levels.

Future Challenges

Arsenic crisis is a global problem and the solution is in development of breakthrough technologies. In this connection, two major areas contributing significantly towards dearsenification includes molecularly engineered materials and biotechnological advancements. Self-assembled monolayers on mesoporous materials (SAMMS) is an award winning technology developed by PNNL, USA for removal of arsenic from water with fast kinetics (0.1 s), enhanced adsorption and excellent selectivity which needs to be further tested for their field applications. Another field of interest is the use of genetic engineering to create plants that could clean arsenic from contaminated soil and groundwater. By inserting two bacterial genes into thale cress, Arabidopsis thailana, US researchers have created a plant that not only grows well in the presence of arsenic but also is able to store the toxin in its leaves (Dhankher, 2002).

An enzyme directly responsible for converting arsenite to arsenate has been recently discovered. Efforts are underway to identify the same enzyme in other microbes. Also other proteins and genes involved in consuming arsenate are being explored. These findings are to be used to set up bioremediation system for cleaning up mining waste water and also provide safer drinking water for areas such as Bangladesh and West Bengal.

REFERENCES

Deshpande, L.S. and Pande, S.P., 2005. Development of Arsenic Testing Field Kit—A Tool for Rapid On-site Screening of Arsenic Contaminated Water Sources, Environmental Monitoring and Assessment, 101, 93.

Chakraborti, D., Ahamed, S., Rahman, M.M., Sengupta, M.K., Lodh, D., Das, B., Hossain, A., Mukherjee, S.C., Pati, S. and De, N.K., 2004. Risk of Arsenic Contamination in Groundwater, Environmental Health Perspectives, 112(1), A19-A21.

Daus, B., Wennrich, R. and Weiss, H., 2004. Sorption Materials for Arsenic Removal from Water: Comparative Study, Water Research, 38, 2948–54.

Shevade, S. and Ford, R.G., 2004. Use of Synthetic Zeolites for Arsenate Removal from Pollutant Water, Water Research, 38, 3197-204.

Chakraborti, D., Mukherjee, S.C., Pati, S., Sengupta, M.K., Ramman, M.M., Chowdhury, H.K., Lodh, H., Chanda, C.R., Chakraborti, A.K. and Basu, G.K., 2003. Arsenic Ground Water Contamination in Middle Ganges Plain, Bihar, India: A Future Danger? Environmental Health Perspective, 111(9), 1194-201.

Chakraborti, D., Sengupta, M.K., Rahman, M.M., Chowdhury, U.K., Lodh, D., Chanda, C.R., Basu, G.K., Mukherjee, S.C. and Saha, K.C., 2003. Arsenic Exposure and Health Effects, Edt. Chappell, W.R., Abernathy, C.O., Calderon, R.L., Elsevier Science, New York.

NIEHS, 2003. Arsenic Exposure and Human Health, Research Brief, 104, Super Fund Basic Research Program, National Institute of Environmental Health Sciences.

Sharpe, M., 2003. Deadly Waters Run Deep: The Global Arsenic Crisis, *Journal of Environmental Monitoring*, 5, 81N-85N.

Zeng, L., 2003. A Method for Preparing Silica-containing Iron(III) Oxide Adsorbents for Arsenic Removal, Water Research, 37, 4351-58.

Dhankher, P., Li, Y., Rosen, B.P., Shi, J., Salt, D., Senecoff, J.F., Sashti, N.A. and Meagher, R.B. 2002. Engineering Tolerance and Hyperaccumulation of Arsenic in Plants by Combining Arsenate Reductase and y-glutamylcysteine Synthetase Expression, Nature Biotechnology, 20, 1140-45.

WHO, 2002. Arsenic Mass Poisoning on an Unprecedented Scale, WHO Feature 206, World Health Organisation.

Xu, Y., Nakajima, T. and Ohki, A., 2002. Adsorption and Removal of Arsenic(V) from Drinking Water by Aluminum-loaded Shirasu-zeolite, *Journal of Hazardous Materials*, B92, 275-87.

Elizalde-Gonzalez, M.P., Mattusch, J. and Wennrich, P.M., 2001. Uptake of Arsenite and Arsenate by Clinoplolite-rich Tuffs, Microporous and Mesoporous Materials, 46, 277-86.

Elizalde-Gonzalez, M.P., Mattusch, J., Wennrich, P.M. and Einicke, W.D., 2001. Sorption on Natural Solids for Arsenic Removal, *Chemical Engineering Journal*, 81, 187-95.

Pande, S.P., Deshpande, L.S. and Kaul, S.N., 2001. Laboratory and Field Assessment of Arsenic Testing Field Kits in Bangladesh and West Bengal, India, Environmental Monitoring and Assessment, 68(1), 1.

NEERI, 2000. Study of Arsenic Contamination in the Groundwaters of Block Chowki, District Rajnandgaon (M.P.) for Public Health Engineering Department, Rajnandgaon Division, M.P., April 2000.

Ghurye, G.L., Clifford, D.L. and Tripp, A.R., 1999. Combined Nitrate and Arsenic Removal by Ion Exchange, *Journal of the American Water Works Association*, 91, 85-96.

Lee, C., Low, K., Liew, S. and Choo, C., 1999. Removal of Arsenic (V) from Aqueous Solution by Quaternized Rice Husk, *Environmental Technology*, 20(9), 971-78.

Facing the Challenges of a New Century, 30 Years of CEPIS (1968-1998), 9 Document by PAN American Center for Sanitary Engineering and Environmental Services.

Pande, S.P., Deshpande, L.S., Patni, P.M. and Lutade, S.L., 1997. Arsenic Removal Studies in Some Ground Waters of West Bengal, India, *Journal of Environmental Science and Health*, A32(7), 1981.

Joshi, A. and Chaudhuri, M., 1996. Removal of Arsenic from Ground Water by Iron Oxide-coated Sand, *Journal of Environmental Engineering*, 122(8), 769-72.

Mallevialle, J., Odendaal, P.E. and Wiesner, M.R., 1996. Water Treatment Membrane Processes. McGraw-Hill, New York.

Mandal, B.K., Chowdhury, T.R., Samanta, G., Basu, G.K. and Chowdhury, P.P., 1996. Arsenic in Groundwater in Seven Districts of West Bengal, India—The Biggest Arsenic Calamity in the World. *Current Science*, 70(11), 976-87.

Stumm, W. and Morgan, J.J., 1996. *Aquatic Chemistry: Chemical Equilibria and Rates in Natural Waters*, Wiley, New York.

Cheng, R.C., Liang, S., Wang, H.C. and Beuhler, M.D., 1994. Enhanced Coagulation for Arsenic Removal, *Journal of the American Water Works Association*, 86(9), 79-90.

NEERI, 1988. Determination of Low Levels of Arsenic in Drinking Waters, *Technical Digest* No. 82.

Pande, S.P. and Sarin, R., 1980. Determination of Nanogram Levels of Arsenic in Drinking Water, *Indian Journal of Environmental Health*, 22 (1), 187.

Pande, S.P., 1980. Morpholine as a New Substitute for Pyridine in the Determination of Arsenic in Water, *Journal of Institute of Chemist* (India), 52, 256.

Pande, S.P. and Hasan, M.Z., 1979. New Solvent for SDDC in Determination of Arsenic, *Indian Journal of Environmental Health*, 21, 332.

Pande, S.P. and Hasan, M.Z., 1978. Improved Method for Determination of Arsenic in Water, *Indian Journal of Environmental Health*, 20(2), 121.

NEERI, 1977. Arsenic Determination in Drinking Water, *Technical Digest* No. 56.

Water Pollution and Health Impacts

16

J.S. Pandey

Preamble

Freshwater ecosystems have been critical to sustaining life and establishing civilizations throughout history. Humans rely on fresh water systems not only for drinking water, but also for agriculture, transportation, energy production, industrial processes, waste disposal, and the extraction of fish and other products. As a result, human settlements worldwide are concentrated near fresh water ecosystems, with over half of the world's population living within 20 km of a perennial river (Small and Cohen, 1999). However, fresh water ecosystems all over the world are now threatened not only by over exploitation and poor management but also because of rising environmental pollution (Lau *et al.*, 1998; Schindler *et al.*, 1995; WWF, 2002), dumping of industrial effluent, run-off from agricultural fields and increasing use of synthetic organic substances (http://edugreen.teri.res.in/explore/water/health.htm).

Current Status

Central Pollution Control Board of India has set up several river quality monitoring stations. In 1999, there were 507 such stations of which 430 were set up under the Monitoring of Indian National Aquatic Resources (MINARS) programme, 50 stations under Global Environmental Monitoring Systems (GEMS), and 27 stations under the Yamuna Action Plan (YAP). The major water polluting sources identified were domestic pollution from large cities and industrial pollution mainly from sugar industry, distilleries, tanneries or fertilizers (CPCB, 1999). The parameters used for the comparison were DO, BOD, Total coliforms and

Faecal coliforms. While BOD excess could be the result of both domestic or industrial pollution, coliform count is directly related to domestic waste water (http://www.cerna.ensmp.fr).

Health Impacts of Water Pollution

Water-borne diseases are infectious and they spread either directly or through flies or filth. There are three types of possible infections due to pathogens: bacterial, viral and protozoal. Typhoid, cholera and dysentery are caused due to bacterial infection, while hepatitis and poliomyelitis are due to viral infection. Amoebic dysentery is caused due to protozoal infection. These pathogens are found in the feces of infected persons.

A large number of chemicals that either exist naturally in water or are added through various anthropogenic activities also lead to various diseases. Often times, heavy metals and toxic compounds can get volatilized from warmer parts of the world, enter the atmosphere and re-condense in cooler areas, thereby contaminating even pristine sites and food supplies (Dewailly, 1989; Kidd *et al.*, 1995; Wania and Mackay, 1993). For instance, the organophosphates and the carbonates present in pesticides affect and damage the nervous system and can even cause cancer. Lead can also affect central nervous system, fluoride leads to yellowing of the teeth and damage to the spinal cord, excess of nitrates may lead to 'blue baby'-syndrome (less oxygen to the brain) and cancer of the digestive tract, long exposure to petrochemicals leads to cancer, while arsenic poisoning through water can cause liver or nervous system damage (http:/ /edugreen.teri.res.in/ explore/water/health.htm).

Immediate Concerns

Water supplies are also vulnerable to pollution from other sources related to human production systems such as toxic waste from mining operations. There is no reliable information on the health impact of almost two-thirds of the 70000 chemical products used in US-based industries. Bio-availability from pollution depends upon the form of the chemical in the water body and factors such as water solubility that affect entry into the food-chain. For example, the residues of many first and second-generation pesticides function as so-called endocrine disruptors i.e. chemicals with the potential to affect the harmone system in humans and animals. Risk assessment for water pollution thus requires long-term monitoring, since the harmful effects may take several years to cause damage (http://www.who.int/water_sanitation_health/ industrypollution/ en/print.html).

Emerging Issues

Understanding and quantifying the risk posed by contaminants in water involves looking at the total environment, the physical, chemical, biological and socio-cultural. Exposure to toxic substances can occur through various routes namely, air, food, transportation routes, soil and water. The risk is much greater if the people in the region are already suffering from malnutrition, poverty and poor sanitation.

Future Challenges

The health risks can be defined in terms of either acute or chronic (long-term) exposure. Acute exposure is defined as exposure for less than 24 hours and usually involves a single dose of a chemical. Long-term exposure refers to repeated or continuous exposure for more than 3 months. The health effects differ markedly between acute and chronic exposure. In most cases of contamination via water, there is more concern about long-term effects. For example, in case of chemical-accumulation in the sediments of rivers and reservoirs, or where industrial discharges continue over a long period of time. Hence, the linkages between health and different sectors of environment need to be more clearly understood and quantified.

REFERENCES

WWF, 2002. Living Planet Report 2002.

CPCB, 1999. Water Quality Status and Statistics, Central Pollution Control Board.

Small, C. and Cohen, J.E., 1999. Continental Physiography, Climate and Global Distribution of Human Population. Proceedings of the International Symposium on Digital Earth, Chinese Academy of Science, Beijing, China, 965-71. (http://www.Ideo.columbia.edu/~small/PDF/ISDE_SmallCohen.pdf)

Lau, S., Mohamed, M., Tan Chi Yen, A. and Su'ut, S., 1998. Accumulation of Heavy Metals in Fresh Water Mollusks, Science of the Total Environment, 214, 113-21.

Kidd, K.A., Schindler, D.W., Muir, D.C.G., Lockhart, W.L. and Hesslein, R.H., 1995. High Concentration of Toxaphene in Fishes from a Sub-arctic Lake, Science, 269, 240-42.

Schindler, D.W., Kidd, K.A., Muir, D.C.G. and Lockhart, W.L., 1995. The Effects of Ecosystem Characteristics on Contaminant Distribution in Northern Freshwater Lakes. Science of the Total Environment 160/161: 1-17.

Wania, F. and Mackay, D., 1993. Global Fractionation and Cold Condensation of Low Volatility Organo-chlorine Compounds in Polar Regions, Ambio, 22, 10-18.

Dewailly, E.A., Nantel, J.P., Weber, J.P. and Meyer, F., 1989. High Levels of PCBs in Breastmilk of Women from Arctic Quebec, *Bulletin of Environmental Contamination and Toxicology*, 43, 641-46.

WASTEWATER

Municipal Wastewater—Treatment (Primary and Secondary)

17

T. Nandy

Preamble

Urban environmental management is one of the most pressing issues, as the urbanization trends continue globally. One of the challenges faced by urban planners, engineers and managers is to ensure provision of basic human amenities such as availability of safe water and sanitation. The improper management of wastewater in many urban areas presents a major challenge. Direct discharge of this wastewater into natural water bodies results in their contamination. In addition, the cumulative impact of untreated waste water discharges can have broad degenerative effects on both public health and ecosystem. Therefore, proper treatment and management of municipal wastewater has direct bearing on water quality issues.

Current Status

The inventory of municipal wastewater for major states is shown in Table 1.

International Scenario

The scenario of urban wastewater management in Europe calls for agglomerations with less than 2000 p.e., equipped with wastewater collection systems. They have to adopt appropriate wastewater treatment by the end of year 2005 under European "Urban Wastewater Directive" 91/271/EEC. The term "appropriate" refers to the impact on the receiving water body, i.e. the wastewater treatment level shall be chosen according to the sensitivity of the receiving water body. This means that authorities are open to different solutions at different levels, in

Table 1. Inventory of Municipal Wastewater for Major States

Sl. No.	State	Population[a] (2001)	Water Requirement (2001)[b]		Domestic Wastewater Generated (mld)	
			km[3c]	mld	Based on Water Requirement*	Based on Population**
1.	Andhra Pradesh	7,57,27,541	2.927	8019	6415	6262
2.	Bihar	8,28,78,796	3.237	8868	7095	14736
3.	Delhi	1,37,82,976	0.981	2688	2150	1629
4.	Gujarat	5,05,96,992	2.052	5622	4498	6654
5.	Karnataka	5,27,33,958	2.12	5808	4647	4740
6.	Kerala	3,18,38,619	1.239	3395	2716	3041
7.	Madhya Pradesh	6,03,85,118	2.968	8132	6505	6552
8.	Maharashtra	9,67,52,247	4.044	11079	8864	14324
9.	Orissa	3,67,06,920	1.153	3159	2527	5471
10.	Punjab	2,42,89,296	0.942	2581	2065	2731
11.	Rajasthan	5,64,73,122	1.984	5436	4348	8656
12.	Tamil Nadu	6,21,10,839	2.599	7121	5696	4462
13.	Uttar Pradesh	16,60,52,859	6.094	16696	13357	17863
14.	West Bengal	8,02,21,171	3.087	8458	6766	9060

[a] indiastat.com (Office of the Registrar General and Census Commissioner, India)

[b] indiastat.com (Central Water Commission)

[c] Annual Water requirement

* Water Requirement × 0.8

** Ratio of Population for 2001 and 1991 multiplied by w/w estimated in 1991.

the case of design or upgrading of small plants. Prevailing urban wastewater management practices in some of the European countries are briefed further. In Italy, most plants (57%) are activated sludge systems (AS). Stabilization ponds are not common in Italy, as in many of the other European countries, mainly due to land availability and groundwater pollution possibility. Rotating Biological Contactor (RBC) concepts are considered with interest, due to low energy and maintenance requirements. France has about 7200 small treatment plants of size upto 2500 p.e., out of which 3250 plants are less than 500 p.e. and 5200 plants are less than 1000 p.e. Majority of wastewater treatment plants in France employ primary and secondary level treatment (trickling filters - 14.5% and lagoons - 21.9%). All plants serve about 8 million p.e. but still insufficient to meet increased urban

Table 2. Wastewater Generation, Collection and Treatment in Major Cities

City	Population	Wastewater Generated (mld)	Wastewater Collected %	Treatment Capacity (mld)	Treatment Facility P	Treatment Facility S	Mode of Disposal
Ahmedabad	3312216	520	80	430	Y	Y	Sabarmati river
Bangalore	4130288	375	75	290	Y	Y	Valley
Bhopal	1062771	189.3	50	87	Y	Y	Agriculture
Chennai	5421985	276	93.1	257	Y	Y	Adyar & Cooum rivers, Irrigation
Coimbatore	1100746	60	75	-	-	-	Noyal river, Irrigation
Delhi	8419084	1270	80	981	Y	Y	Agriculture, Yamuna river
Hyderabad	4344437	348.3	80.1	115	Y	-	River, Irrigation
Indore	1109056	145	80	14	Y	-	Khan river, Irrigation
Jaipur	1518235	220	75	27	Y	Y	Agriculture
Kanpur	2029889	200	75	41	Y	Y	Ganga river, Sewage farm
Kochi	1140605	75	60	-	-	-	Cochin Back waters
Kolkata	11021918	1384	75.1	-	-	-	Hoogly river/ Fish farm
Lucknow	1669204	106	75.5	-	-	-	Gomati river
Ludhiana	1042740	94.4	49.8	-	-	-	Agriculture
Madurai	1085914	48	70	-	-	-	Agriculture
Mumbai	12596243	2228	90	109	Y	Y	Sea
Nagpur	1664006	204.8	79.6	45	Y	Y	Agriculture
Patna	1099647	219	74.9	105	Y	N	River, Fisheries
Pune	2493987	432	85	170	Y	Y	River
Surat	1518950	140	80	70	Y	-	Garden Creek
Vadodara	1126824	120	75	81	Y	Y	River, Agriculture
Varanasi	1030863	170	74.7	101	Y	Y	Ganga river, Agriculture
Vishakha-patnam	1057118	68	80.9	-	-	-	-
Total	70996726	8893	80.6	2923	-	-	-

wastewater treatment requirements. Lagoon concepts are popular in France as their running costs are only 20% of a conventional AS system. Process efficiency and nitrification/denitrification favours AS system, but these have drawbacks related to high running costs, sensitive operation management, significant sludge volumes and problems and a large impact of hydraulic inlet variations, often due to poor sewer systems. People's interest in trickling filters is getting lost in France. There is an increasing interest on chemical treatment units, particularly in the coastal and touristic mountainous areas, but these plants are normally larger ones. Norway mainly has septic tank systems, a few pond systems, and sewer systems with simple mechanical treatment, in the rural areas upto 1970. Chemical plants are much used in Norway, and they are among the best in the world, even small plants have BOD efficiencies more or less equal to those of biological systems.

Public Health and Safety

The most critical issue concerning reclaimed wastewater relate to the protection of public health. Some negative effects of municipal wastes have been documented for over a century, and some of these issues remain relevant even to this date. These include pathogens, oxygen depletion, eutrophication, toxicity of effluents, bio-solids/sludges, intermittent discharges, byproducts of treatment, pharmaceuticals and personal care products, etc. In Canada, many beaches in urban areas are frequently closed for several days during rainfall events because of microbial contamination caused by MWTP overflows. In 1992, approximately 3018 sq km of beach area was closed to shellfish harvesting due to bacterial contamination along the three Canadian coasts (Chambers *et al.*, 1997). Isolated incidences of microbial contamination of drinking water sources in Canada from inadequately treated MWTP effluents have also been reported.

New issues continue to become apparent as the science of understanding the environment advances, and the tools and technology of environmental monitoring develop. MWTP effluents have traditionally been regulated based on specified level of technology or end-of-the-pipe effluent criteria. This neither guarantees environmental protection nor ensures compliance with all environmental regulations. Ecological Risk Assessment (ERA) is used to manage municipal liquid wastes, which addresses the site specific constraints of different receiving environments. However, ERA frequently requires extensive environmental information and interpretation skills not routinely possessed by municipal waste collection and treatment operators. Guidance and development of environmental tools and regulatory framework for municipalities to conduct ERA is needed. This approach would have to ensure that the recommendations are based on unambiguous science and provide regulatory clarity for municipalities. In addition, unlike fresh water irrigation, use of reclaimed wastewater for irrigation is restricted to certain uses due to public health or water quality concerns (Asano *et al.*, 1996; Mills and Asano, 1996). The renovation and reuse of wastewater without compromising on public health are the prime issues that need to be addressed.

Immediate Concern

Formulation of a control algorithm that should include:

- Foolproof wastewater collection and capacity augmentation of existing treatment facilities.
- Appropriate wastewater treatment to reduce concentration of microorganisms to meet the statutory norms / WHO (1989) guidelines.
- Application methods (irrigation) for reduction/ minimization of contact of wastewater with edible crops.
- Human exposure control for workers, crop-handlers and ultimate consumers (WHO, 1989; Mara and Cairncross, 1989; Strauss and Blumenthal, 1990).
- Water conservation efforts.

Emerging Issues

Ecological engineering is emerging as a field with potential to conserve the natural environment and at the same time adapting to solving environmental pollution problems which are sometimes intractable (Mitsch and Jorgensen, 1989). Todd and Josephson (1996) have stated that ecological engineering will influence the future of waste treatment, environmental restoration and remediation, food production and the design of human settlements. Wang *et al.* (1998) and Qixing *et al.* (1996) have stated that the systematic planning of wastewater reuse schemes employing a combination of technologies for food production, may offer one solution to solving food shortages and water pollution.

The development of solar technologies and an increased understanding of the role of organisms in the water purification process are providing both economic and environmental benefits (Todd and Todd, 1994). It provides ecologically based "green machines" or "living machines" (Guterstam and Todd, 1990; Mitsch, 1991) that capture the same natural mechanisms occurring in natural wetland treatment systems and treat wastewater in confined space environments, and are therefore, suitable for densely populated urban areas. In these systems, enclosed greenhouses enhance the growth of algae, plants and bacteria which in turn act to degrade the biological and pathogenic components of the wastewater effluent. Recovered treated wastewater from these systems can be used for landscape irrigation, and for the propagation of horticultural plants for resale (Farrell, 1996; WEF, 1995).

Adoption of disinfection technologies based on scientific approach prior to land application of treated domestic wastewater is also an issue of concern from public health viewpoint.

Management and Control

Control of the urbanization phenomena is the problem associated with providing municipal services and water sector infrastructure, including the provision of both fresh water resources and sanitation services. Providing housing, health care, social services, and access to basic human needs (infrastructure such as clean water and the disposal of effluent), present major challenges to engineers, planners and politicians (Black, 1994; Giles and Brown, 1997).

In developing countries, 300 million urban residents have no access to sanitation and it is mainly low-income urban dwellers that are affected by lack of sanitation infrastructure (Forget, 1992; Briscoe and Steer, 1993; Black, 1994; Veenestra and Alaerts, 1996; Giles and Brown, 1997). Approximately two-thirds of the populations in the developing world have no hygienic means for disposal of excreta and even lack adequate means for disposal of wastewater (Sinnatamby, 1990; Niemczynowicz, 1996).

Future Challenges

Preserving water resources and the provision of sanitation amidst rapid urbanization and industrialization in many urban centres of the developing world are the major future challenges to be attended. In India, like many developing nations, planning for domestic wastewater reuse is one area that has not received adequate attention; and to compound the problem, many existing treatment facilities are in poor shape (Chawathe and Kantawala, 1987). Development of natural systems for treatment of domestic wastewater at low operation and maintenance costs including onsite demonstration of such technologies and training of O&M staff are the urgent requirements under the prevailing conditions.

REFERENCES

Wang, R., Yan, J., Lu, B. and Hu, D., 1998. The Practice of Integrated Bio-Systems in China, Foo, E.L. and Della Senta, T. (Eds), International Conference on Integrated Bio-systems (ICIBS), 1-11. (http://home2.swipnet.se/w 25860/icibs/wang/)

Chambers, P.A., Allard, M., Walker, S.L., Marsalek, J., Lawrence, J., Servos, M., Busnarda, J., Munger, K.S., Adare, K., Jefferson, C., Kent, R.A. and Wong, M.P., 1997. The Impacts of Municipal Wastewater Effluents on Canadian Waters: A Review, *Water Quality Research Journal of Canada*, 32, 659-713.

Giles, H. and Brown, B., 1997. And Not a Drop to Drink, Water and Sanitation Services in the Developing World, *Geography*, 82 (2), 97-109.

Asano, T., Maeda, M. and Takaki, M., 1996. Wastewater Reclamation and Reuse in Japan: Overview and Implementation Examples, Water, Science and Technology, 34 (11), 219-26.

Farrell, M., 1996. Purifying Wastewater in Green Houses, BioCycle, January 1996.

Niemczynowicz, J., 1996. Megacities from a Water Perspective, Water International, 21(4), 198-205.

Qixing, Z., Limei, D. and Bell, R.W., 1996. An Integrated Plan for Town-enterprise Reuse and Wetland Strategy: A Case Study, *Desalination*, 106, 439-42.

Todd, J. and Josephson, B., 1996. The Design of Living Technologies for Waste Treatment, Ecological Engineering, 6, 109-36.

Veenestra, S. and Alaerts, G., 1996. Technology Selection for Pollution Control. A. Balkema, H. Aalbers and E. Heijndermans (Eds.), Workshop on Sustainable Municipal Waste Water Treatment Systems, Leusdan, the Netherlands, ETC in co-operation with WASTE. 12-14 November 1996, 17-40.

Helge, Brattebo, 1995. *Journal of World Water*, University of Trondheim.

WEF (Water Environment Federation), 1995 Greenhouse Filter System Yields Zero Discharge. *Water Environment and Technology*, June 1995.

Black, M., 1994. Mega-slums: The coming sanitary crisis. London: Water Aid.

Todd, J. and Todd, N., 1994. From Eco-cities to Living Machines; Principles of Ecological Design. Berkeley, CA: North Atlantic Books.

Briscoe, J. and Steer, A., 1993. New Approaches to Sanitation—A Process of Structural Learning. Ambio, 22 (7), 456-59.

Forget, G., 1992. Health and the Environment: A People-centered Strategy, Ottawa: International Development and Research Centre.

Mitsch, W.L., 1991. Ecological Engineering: The Roots and Rationale of a New Ecological Paradigm, C. Etnier and B. Guterstam (Eds.), Ecological Engineering for Waste Water Treatment, Gothenburg, Sweden: Bokskogen, 110-20.

Guterstam, B. and Todd, 1990. Ecological Engineering for Wastewater Treatment and its Application in New England and Sweden, Ambio, 19, 173-75.

Sinnatamby, G., 1990. Low Cost Sanitation, J. Hardroy, S. Cairncross and D. Satterthwaite (Eds.), The Poor Die Young: Housing and Health in the Third World Cities, London: Earthscan Publications.

Mara, D. and Cairncross, S., 1989. Guidelines for the Safe use of Wastewater and Excreta in Agriculture and Aquaculture, Geneva: United Nations Environmental Programme/World Health Organization.

Chawathe, S.D. and Kantawala, D., 1987. Reuse of Water in City Planning, Water Supply, 15 (1), 17-23.

Mitsch, W.J. and Jorgensen, S., 1989. Introduction to Ecological Engineering, W.J. Mitsch and S. Jorgensen (Eds.), Ecological Engineering: An Introduction to Ecotechnology, New York: J. Wiley and Sons, Inc., 3-12.

Industrial Wastewater

18

N.N. Rao

Preamble

The production and consumption of a product by mankind leads to pollution, and it directly impacts the quality of air, water and soil. The impact is further extended to all living beings that are dependent for their survival on any one or all of these receptors. Therefore, reduction/mitigation of industrial pollution is important.

Current Status

The estimates indicate that in about 8432 large and medium scale polluting industries are in India (CPCB, 1997). However, small-scale industries in India are estimated to be around 3.2 lakhs (Maria, 2003) and the number cannot be ascertained exactly as many of the units are non-registered and operate in the unorganized sector.

There has been an increase in water demand for industrial needs leading to enhanced wastewater generation. The total quantity of wastewater generated from various industries can be arrived at based on the wastewater generation with respect to the specific industry as described in Table 1. The contribution (%) of wastewater generated from major industries in India is presented in Fig. 1, and the quantity (mld) generated from major small scale industries is presented in Table 2.

All medium and large-scale industries in India possess independent effluent treatment plants (ETPs) comprising at least primary and secondary treatment facilities. Some selected industries also focus on tertiary treatment. The pollutant

Table 1. General Wastewater Standards for Discharge of Industrial Effluent

Industry	Quantity Per Unit of Product
Integrated Iron & Steel	16 m^3/tonne of finished steel
Sugar	0.4 m^3/tonne of cane crushed
Pulp & Paper Industries:	
Large Pulp & Paper:	
- Pulp & Paper	175 m^3/tonne of paper processed
- Rayon grade Pulp	150 m^3/tonne of paper
Small Pulp & Paper:	
- Agro-residue based	150 m^3/tonne of paper processed
- Waste paper based	50 m^3/tonne of paper processed
Fermentation Industries:	
- Malt	3.5 m^3/tonne of grain processed
- Brewery	0.25 m^3/KL of beer produced
- Distillery	12 m^3/KL of alcohol produced
Caustic Soda:	
- Membrane cell process	1 m^3/ tonne of caustic soda produced excluding cooling tower blowdown
- Mercury cell process	4 m^3/tonne of caustic soda produced (mercury bearing). 10% blowdown permitted for cooling tower.
Textile Industries: Man-made fibre:	
- Nylon & Polyster	120 m^3/tonne of fibre produced
- Viscose Staple Fibre	150 m^3/tonne of product
- Viscose Filament Yarn	500 m^3/tonne of product
Tanneries	25 m^3/tonne or raw hide
Starch Glucose and related products	8 m^3/ tonne of maize crushed
Dairy	3 m^3/tonne of Milk
Natural rubber processing industry	4 m^3/tonne of rubber
Fertiliser:	
- Straight nitrogenous fertiliser	5 m^3/tonne of urea or equivalent produced
Straight phosphatic fertiliser excluding manufacture of any acid	0.5 m^3/tonne of fertilizer

Source: Environmental Standards for Ambient Air, Automobiles, Fuels, Industries and Noise, Central Pollution Control Board, Ministry of Environment & Forest, New Delhi.

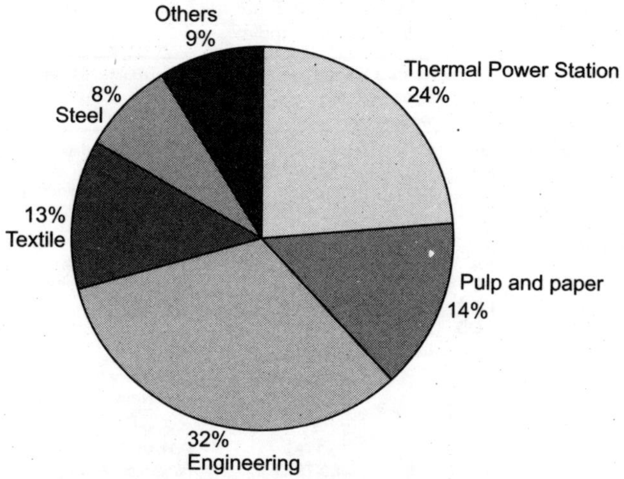

Fig. 1. Percent Contribution of Wastewater by Major Industries in India (Maria, A., 2003)

Table 2. Wastewater Generation by SSIs in Select Industrial Sectors

Industry	Wastewater Generation (mld)
Engineering	2125
Paper and board mills	1087
Textile	450
Organic chemicals	60
Tanneries	50
Pharmaceuticals	40
Dye and dye intermediates	32
Soaps, paints, varnishes, and petrochemicals	10
Edible oil and vanaspati	7

Source: CPCB, 1997.

characteristics and the general treatment of some of the major industries in India are presented in Table 3. By and large the small-scale units located in clusters have common effluent treatment plants (CETPs). Some of the details of operational CETPs in India are presented in Table 4.

Table 3. Pollution Characteristics and Treatment Schemes for Major Industries in India

Sl. No.	Type of Industry	Characteristics Range (All values in mg/l except pH)		Treatment Scheme Adopted
1.	Pulp & Paper	pH	6.9-9.8	- Primary (Chemical Aided)
		SS	1280-21000	- Secondary
		BOD	150-420	(Anaerobic & Aerobic)
		COD	700-1000	
2.	Sugar Mill	pH	4.2-6.0	- Primary
		SS	400-860	- Secondary
		TS	660-1490	(Anaerobic & Aerobic)
		BOD	185-245	
		COD	1080-1710	
3.	Dairy	BOD	2790-6760	- Primary
		COD	4100-10800	- Secondary
		O & G	1350-2180	(Anaerobic & Aerobic)
4.	Textile	pH	2.0-11.9	- Segregation of Dye Bath
		SS	70-1725	- Primary Treatment
		TDS	460-7560	(Chemical Aided)
		TS	530-6400	- Secondary
		BOD	80-1100	(Aerobic)
		COD	248-3390	- Tertiary Treatment
		Chlorides	80-1100	(Adsorption)
		Sulphate	70-1600	
		Sodium	1010	
5.	Tannery	pH	9.4-9.7	- Segregation of Soak
		SS	1200-6000	Liquor Solar Pans
		TDS	9000-13000	- Primary (Chemical Aided)
		BOD	900-1200	- Secondary Treatment
		COD	1800-2000	(Aerobic & Anaerobic)
		Chromium	5-26	- Tertiary (Chemical Aided)
		Sulphate	1-60	
6.	Refinery	pH	6.5-9.0	- Primary (Chemical Aided)
		Free Oil	2000-3000	- Secondary
		Emulsified Oil	80-120	(Aerobic & Anaerobic)
		SS	200-400	- Tertiary
		BOD	100-300	(Filtration & Adsorption)
	Phenolic	12-30		
		Compounds		
		Sulphides	10-220	
7.	Distillery	pH	3.5-4.5	- Primary (Anaerobic)
		SS	4400-6400	- Secondary (Aerobic)
		TDS	90050-101600	- Biocomposting
		TS	94450-108000	- Land Application
		BOD	29400-30000	
		COD	103000-107000	
		Total Nitrogen	800-1200	
		Total Phosphate	30-390	
		Sodium	150-200	
		Potassium	500-600	
		Sulphate	1450-2500	
		Sulphides	320-360	

Source: Recent Advances in Wastewater Engineering, Book Enclave, Jaipur, 2004.

Table 4. Some Operational CETPs in India

Location	Treatment Capacity (mld)	Member Units	Member Type	Capital Cost (Rs. Million)	Operating Agency
Andhra Pradesh					
Jeedimetla	1.50	109	Drugs & Chemicals	25.0	Jeedimetla ETP Ltd.
Pattencheru	7.50	72	Drugs & Chemicals	62.4	Pattancheru ETP Ltd.
Bolarlem	0.25	26	Drugs & Chemicals	2.50	Progressive ETP Ltd.
Delhi					
Wazirpur Ind. Area	24.00	686	Engg., Pickling, Dying, Printing	287	
Mangolpuri Ind. Area	2.40	92	Engg., Metal Finishing, Food	65.0	Delhi State Industrial Development Plastics Corporation, Delhi
Mayapuri Ind. Area	12.00	667	Engg., Pickling, Chemical, Printing	134.0	
Jhilmil Ind. Area	16.80	139	Engg., Electroplating, Chemical	153.0	
Lawrence Rd. Ind. Area	12.00	251	Engg., Dyeing, Plastic, Food processing	109.0	
GT Karnal Rd. Ind. Area	6.00	415	Engg., & Pickling	99.0	
SMA Ind. Area	12.00	39	Engg., & Food	217.0	
Nangloi Ind. Area	12.00		Engg., Pickling, Pharma., Textiles	159.0	
Gujarat					
Vatva, Ahmedabad	16.00	437	Textile, Distillery, Pharmaceuticals	300.0	The Green Env. Services Co. Ltd. Chemicals
Odhav, Ahmedabad	1.20	57	Dye & Dye Intermediates	41.5	Odhav Enviro Projects Ltd.
Ankleshwar	1.00	200	Dye & Dye Intermediates, Textiles, Pharmaceuticals, Pesticides, Pigments	68.0	Enviro Technology Ltd., Ankleshwar
Vapi	55.00	615	Chemicals, Plastics, Paper Pharmaceuticals, Garments, Engg.	204.0	Vapi Industries Association
Jetpur, Rajkot	20.00	2000	Dyeing & Printing	50.0	Jetpur Dyeing & Printing Assocn.
Nandesari, Vadodara	5	–	Chemicals, Dyes & Dye Intermediates	–	Nandesari Industrial Association
Sarigam, Valsad	0.40	7	Chemicals	–	Perfect Enviro. Control System Ltd., GIDC, Sarigam
Gumsav, Odhav, Ahmedabad	1.00	–	–	15.0	Gujarat Vepari Mahamandal S.A.V. Ltd., Ahmedabad
Dharmeshwar, Jetpur, Rajkot	0.15	19	Pharmaceutical	–	Dhareshwar GIDC Vistar Asscn., Rajkot
Panoli Bharuch	1.00	101	–	33.5	Panoli Enviro Technology Ltd., Bharuch

(*Contd*...)

Location	Treatment Capacity (mld)	Member Units	Member Type	Capital Cost (Rs. Million)	Operating Agency
Sachin, Dist: Surat	0.5	35	–	50.0	Globe Enviro Care Ltd., Sachin, Dist. Surat
Sanand, Dist. Ahmedabad	0.2	27	Textile	32.0	Sanand Eco Projects Ltd., Sanand, Dist. Ahmedabad.
Odhav, Ahmedabad	1.00	3	Chemicals	4.50	Odhav Green Enviro Project Association, Odhav, Ahmedabad
Naroda, Ahmedabad	3.00	242	Chemicals	95.0	Naroda Enviro Projects Ltd., Naroda, Ahmedabad
Padra, Dist-Vadodara	2.25	33	Chemicals	64.0	Enviro Infrastructure Co. Ltd., Tal: Padra, Dist. Vadodara
Veraval	0.50	46	Chemicals	77.1	The Veraval Industries Veraval
Gandhinagar	0.08	22	Chemicals	3.00	Kalol GIDC Industrial Association, Dist. Gandhinagar
Narol, Ahmedabad	0.1	13	Chemicals	1.50	Narol Dyestuff Enviro Society, Narol, Ahmedabad
Karnataka					
Kadugondanahalli Bangalore	–	14	Tannery	16.0	Lidkar Enviro Control System Ltd.
Kumbalgod	–	40	Tannery	11.3	Pai & Pai Chemicals (I) Ltd.
Maharashtra					
Thane-Belapur (Navi Mumbai)	12.00	400	Dyes & Dye Intermediates, Pharmaceuticals, Chemicals, Drugs, Drug & Intermediates	45.0	CETP (T-B) Assn., MIDC, Navi-Mumbai
Dombivilli	1.50	331	Dyes & Dye Intermediates, Pharmaceuticals, Chemicals, Drugs, Drug & Intermediates	26	Dombivilli CETP MIDC, Phase-II, Dombivilli (East)
Tarapur, Boisar	1.00	208	Chemicals, Pharmaceuticals, Dyes & Dye Intermediates, Paint & Textiles	33	Tima Co-Op. Society Ltd., MIDC, Boisar, Thane
Punjab					
Jalandhar	1.5	29	Tanneries	9.6	Punjab Small Scale Leather Export Corpn., Jalandhar
Rajasthan					
Pali Unit-1	6.75	–	Textile Units	–	–
Pali Unit-2	6.75	–	Textile Units	–	–

(*Contd...*)

Location	Treatment Capacity (mld)	Member Units	Member Type	Capital Cost (Rs. Million)	Operating Agency
Balotra	6.00	425	Textile Units	30.0	Balotra Water Pollution Control & Res. Foundation Trust
Ranipet	2.5	87	Tannery	29.5	Ranipet SIDCO
Ayyampeettai	–	–	Tannery	–	Ayyampet Muthialpet
Ranipet	4.0	76	Tannery	–	TALCO-Ranipet
Pallavaram	3.0	138	Tannery	70.0	Pallavaram Tanners Indl. Co-op Ltd.
Salem	–	64	Medical	0.45	Salem Medical Waste Management Co. Ltd.
Tamil Nadu					
Karur	2.1	–	Textile Units	–	Karur, Thiruvai
Karur	–	–	Textile Units	–	Amravathi Pollutech
Karur	1.3	–	Textile Units	–	Karur-Karuppampalayam
Melivishram	3.4	36	Tannery	–	Vishram Tanners
Dindigul	2.5	36	Tannery	–	TALCO-Dindigul
Madhavaram	0.8	14	Tannery	–	Madhvaram Leathers
Malligaithope	1.1	18	Tannery	–	Ambur Tannery
Pernambut	0.89	18	Tannery	–	TALCO Pernambut
Thirupur	–	–	Textile Units	–	Andipalyam CETP
Thirupur	–	–	Textile Units	–	Mannarai CETP
Thirupur	–	–	Textile Units	–	Manickpuram CETP
Thirupur	–	–	Textile Units	–	Angeri Palyam CETP
Thirupur	–	–	Textile Units	–	Chinnakarai Thirupur
Thirupur	–	–	Textile Units	–	Kashipalyam CETP
Thirupur	–	–	Textile Units	–	Kunnangalpalayam CETP
Veerapandi	–	–	Textile Units	–	Veerapandi CETP
Bakkalapalli	–	–	Tannery	–	TALCO Perinambut
Uttar Pradesh					
Jajmau, Kanpur	36.00	354	Tannery	220.8	U.P. Jal Nigam
Unnao	2.15	21	Tannery	19.5	UTPCCL, Unnao
Mathura	6.25	30	Textile Units	–	Mathura A.K.P.N. Co. Ltd.

Source
- Performance Evaluation of Common Effluent Treatment Plants, Ministry of Environment & Forest, Govt. of India.
- www. Cpcb.delhi.nic.in/ar2003/ar2-3ch14.htm
- NEERI Report, Phase I Report, Wastewater Management in Cluster of Tanneries in Tamil Nadu, 1997.
- NEERI Report, Phase I Report, Design of Common Effluent Treatment Plants for Industrial Estates in Delhi, 1996.

The various treatment combinations utilized for meeting the regulatory norms are listed hereunder:

Treatment Technologies Available for Wastewater Treatment

Degree of Treatment	Unit Operation/Process
Primary Treatment	• Chemical coagulation and flocculation • Clarifiers • Tube settlers • Sedimentation Tank
Secondary Treatment	**Anaerobic Biological Treatment** • Packed Bed Reactors • Rotating Biological Contactor • Upflow Anaerobic Sludge Blanket (UASB) • Fluidized Bed Reactor **Aerobic Biological Treatment** • Sequential Batch Reactor • Activated Sludge Process • Waste Stabilization Pond • Aerated Lagoon • Oxidation Ditch • Contact Stabilization
Tertiary Treatment	**Filtration** • Sand and Gravel **Adsorption** • Activated Carbon **Advanced Oxidation Process** • Fenton's Reaction • Ozonation • Wet Air Oxidation
Advanced Treatment	**Membrane Techniques** • Ion Exchange • Ultrafiltration • Nanofiltration • Reverse Osmosis • Electrodialysis

More recently, a few industries due to non-availability of fresh water reserves, increased water cess and curtailed water supplies are realizing the need for reduction in water demand. So there is a trend seen to cut water demand through recycle and reuse. Many more industries need to realize the need for substantial reductions in water demand.

Although environmental awareness is prominent in industries' management, however, due to lack of cash returns from such treatment facilities, the commitment is rather half-hearted. This mindset is slowly changing because of the perceivable benefits of recycle and reuse of treated wastewater and other waste byproducts.

International Status

EPA has established effluent discharge guidelines and pretreatment standards for various industries. The industries in developed countries have accumulated much expertise and experience in wastewater treatment process design and operation to comply with the regulatory standards. The industry shares information about the state-of-the-art techniques for addressing their water pollution control activities. They include improvement in traditional treatment facilities as well as process changes and best management practices that lead to reduction in pollutants generation.

Most Serious Concerns

- Small-scale industrial sector where the pollution abatement has been neglected in general.
- Absence of water audit by industries.
- Water resource management, viz., lack of framework regulations on groundwater abstraction and unsustainable groundwater withdrawals.
- Low-key water conservation efforts.
- Lack of willingness to change in-plant operations/machinery. Process technology and machinery need to be revamped so as to minimize generation of wastes.
- Lack of fact sheets on the use of alternative raw materials/reagents that are environmentally benign in the production as well as waste treatment facilities.
- Non-optimal performance of the ETPs and CETPs.

Emerging Issues

- Adoption of clean technologies.
- Appropriate wastewater management practices with recourse to zero discharge, recycle and reuse.

- Intensification of water conservation efforts by industries through regular water audit and resource management approaches.
- Cost effective tertiary level innovative treatment technologies to help achieve recycling and reuse of treated wastewater.
- Private sector participation for wastewater collection and treatment.
- Resource and value recovery.
- VOC emissions from effluent treatment plants and their control.
- Application of membrane techniques for TDS control.
- A sound protocol based on scientific approach for land application of treated effluent.

Future Challenges

- Novel R&D aiming at environment-friendly and cost-effective treatment technologies for wastewater with recourse to resource recovery.
- Developing knowledge base about natural systems that may be used for management and treatment of industrial wastewater.
- Developing technologies for specific pollutant reductions viz., color, toxic metals, recalcitrant organics, POPs.
- Training programmes and on-site demonstrations on wastewater management for effective implementation of preventive and corrective measures.

REFERENCES

A. Maria, 2003. The Costs of Water Pollution in India, Conference on Market Development of Water and Waste Technologies through Environmental Economics, 30th-31st October 2003, Delhi.

CPCB, 1997. National Inventory of Large and Medium Industry and Status of Effluent Treatment and Emission Control System, CPCB.

Microbial Treatment

19

H.J. Purohit and Atya Kapley

Preamble

Microbes with their catabolic diversity are the key players in biogeochemical cycles. They possess enormous metabolic abilities that can degrade or mitigate all sorts of wastes from oil spills to pesticide runoff or sewage waste. By studying the kinetic parameters, the strategies have been devised to improve the biological wastewater treatment methods. However, the "Black Box" that represents the microbial population, which decides the kinetics, has not been properly studied yet. With the advent of molecular tools, the diversity associated with these systems are now being analyzed (Ka *et al.*, 1994; Kapley and Purohit, 2001; Moharikar *et al.*, 2003).

Microbes can biodegrade or transform any chemical, given the right environmental conditions and catabolic system. Many environmental technologies prefer to rely on microorganisms, since they are less expensive than traditional physical/chemical treatment methods. The second major advantage of microbiological processes is that they are flexible; they adapt to variable conditions (self-regulation) and also to new molecules or combinations of new molecules (Grommen and Verstraete, 2002). Hence, there is an interest in the area of microbial populations and functioning of effluent treatment plants.

Current Status

Evidence of wastewater treatment dates far back in history. In ecosystems, organic molecules are either mineralized to CO_2 or are transformed before they can be catabolized to enter the biogeochemical cycles. There are numerous reports on isolation of bacteria capable of degrading pollutants (Kutty *et al.*, 2000; Kutty *et al.*, 2001; Qureshi *et al.*, 2001; Qureshi and Purohit, 2002). However, no single organism possess the catabolic potential to degrade all types of pollutants. Therefore, a mixed culture or consortia approach is more practical in treatment processes (Chhatre *et al.*, 1996; Sharma *et al.*, 2002; Rao *et al.*, 2004; Khan and Anjaneyulu, 2005). Other than bioaugmentation, various methods of biological treatment in effluent treatment plants (ETPs) like biostimulation by addition of nutrients, role of engineering parameters, etc. had been reviewed in Moharikar *et al.* (2005).

Problems in Microbial Treatment

The total quantity of water used by the domestic and industrial sectors is much less than that consumed for irrigation. However, the wastewater from these two sectors has a greater concentration of pollutants and is a greater source of worry. The potential to utilize pollutants in the effluents as carbon sources for metabolism varies enormously among microorganisms (Van der Meer *et al.*, 1992; Cavalca *et al.*, 2000). Moreover, the behaviour of pure isolates in a community requires more exploration. The biomass is very sensitive to sudden shock load of influents, a scenario usually observed in chemical manufacturing units. These factors are largely responsible for the failure of treatment plants (Hall and Melcer, 1983; Ren, 2004).

Immediate Concern

Water recycling is the emerging issue in wastewater management because of the scarcity of water, which is being reported in many countries, where the demand is 2000 m^3 of fresh water per capita per year (Seckler *et al.*, 1999). Other than trace chemicals that contaminate the drinking water supply, there is also an increasing problem of bacterial contamination (Kapley *et al.*, 2000, 2001; Purohit and Kapley, 2002, 2005; Dewettinck *et al.*, 2001).

Emerging Issues

The requirements for treated wastewater are becoming increasingly more stringent, and therefore, the improved efficiency of biological treatment processes is indispensable at industrial effluent treatment plants. Improvements in treatment methods are being looked into. Different bioremediation protocols to enhance decomposition are being evaluated and analyzed using mathematical tools to achieve data reduction for modeling purposes (Gallizia *et al.*, 2005). Novel granules-based bioreactors devising a unified interpretation of the formation of anaerobic and aerobic granules under various operating conditions have been reported. Biogranule-associated bacteria live in a confined space and the advantage of the structured microbial granules is their ability to acquire transmissible, genetic elements at accelerated rates (Liu, Y. and Tay, J-H., 2004). The use of molecular tools along with microbiological aspects of the biomass will greatly enhance the efficacy of the treatment system.

Inadequacy in Management and Control

The control of activated sludge process using the classical control tools is often difficult because of strong non-linearity and time-variant parameters. The Roadblocks to the implementation of bio-treatment strategies has been discussed by Talley and Sleeper (1997). They suggest that the main drawback comes because the scale of the experiment varies from microscale in the laboratory to macroscale at the treatment site. The use of fuzzy logic in controlling such processes has been proved very effective in different studies (Chen *et al.*, 2003; Cadet *et al.*, 2004). Parameters like DO, NH_4 concentration, etc. could be defined linguistically and the fuzzy numbers could be generated for each linguistic term. A fuzzy rule-based system could be developed that incorporates the knowledge about the role of different parameters in deciding the assimilative capacity of the treatment plant (Purohit *et al.*, 2005).

Future Challenges

Xenobiotics biodegradation in the environment occurs in the context of various interactions among microorganisms. To understand this ecological process, identification of functionally important populations is considered to be the primary step. Laboratory studies can then proceed to the analysis of *in situ* activity and interactions with other organisms. Such studies will shape a deeper understanding of the ecology of pollutant degradation and facilitate the development of new bioremediation strategies. Concern about environmental issues such as global climate change, depletion of water, loss of biodiversity, etc., requires the foray of research into the development of more environmentally friendly technologies and energy sources. Enzymes also help to maintain an unpolluted environment through their use in waste management. Recombinant DNA technology, protein engineering, and rational enzyme design are the emerging areas of research pertaining to environmental applications of enzymes. The future will also see the employment of various technologies including gene shuffling, high throughput screening, and nanotechnology (Ahuja *et al.*, 2004).

REFERENCES

Gallizia, I., Vezzulli, L., and Fabiano, M., 2005. Evaluation of Different Bioremediation Protocols to Enhance Decomposition of Organic Polymers in Harbour Sediments, Biodegradation, 16, 569-79.

Khan, Z. and Anjaneyulu, Y., 2005. Influence of Soil Components on Adsorption-desorption of Hazardous Organics-development of Low Cost Technology for Reclamation of Hazardous Waste Dumpsites, *Journal of Hazard Matter*, 14, 161-69.

Moharikar, A., Purohit, H.J. and Kumar, R., 2005. Microbial Population Dynamics at Effluent Treatment Plants, *Journal of Environmental Monitoring*, 7, 552-58.

Purohit, H.J., Kapley, A. and Raje, D.V., 2005. Dyanamics of Microbial Community: An Environmental Perspective, Microbial Diversity: Current Perspective and Potential Applications, T. Satyanarayana and B.N. Johari, Kluwer Publishers, UK, 559-72.

Ahuja, S.K., Ferreira, G.M. and Moreira, A.R., 2004. Utilization of Enzymes for Environmental Applications, Critical Reviews in Biotechnology, 24, 125-54.

Cadet, C., Beteau, J.F. and Hernandez, S.C., 2004. Control Engineering Practice, 12, 335-47.

Liu, Y. and Tay, J-H., 2004. State of the art of Biogranulation Technology for Wastewater Treatment, Biotechnology Advances, 22, 533-63.

Rao, A.G., Naidu, G.V., Prasad, K.K., Rao, N.C., Mohan, S.V., Jetty, A. and Sarma, P.N., 2004. Anaerobic Treatment of Wastewater with High Suspended Solids from a Bulk Drug Industry using Fixed Film Reactor (AFFR), Bioresource Technology, 93, 241-47.

Ren, S., 2004. Assessing Wastewater Toxicity to Activated Sludge: Recent Research and Developments, *Environment International*, 30, 1151-64.

Chen, W.C., Chang, Ni-Bin and Chen, J-C., 2003. Rough Set-based Hybrid Huzzy-neural Controller Design for Industrial Wastewater Treatment, Water Research, 37, 95-107.

Moharikar, A., Kapley, A. and Purohit, H.J., 2003. Detection of Dioxygenase Genes Present in Various Activated Sludge, Environment Science and Pollution Research, 10, 373-76.

Grommen, R. and Verstraete, W., 2002. Environmental Biotechnology: The Ongoing Quest, *Journal of Biotechnology*, 98, 113-23.

Purohit, H.J. and Kapley, A., 2002. Microbial Quality Control of Drinking Water: PCR as an Emerging Option, Trends in Biotechnology, 20, 325-26.

Qureshi, A. and Purohit, H.J., 2002. Isolation of Bacterial Consortium for Degradation of p-nitrophenol from Agricultural Soil. *Annals of Applied Biology*, 140, 159-162.

Sharma, G., Mehra, N.K. and Kumar, R., 2002. Biodegradation of Wastewater of Najafgarh Drain, Delhi using Autochthonous Microbial Consortia: A Laboratory Study, *Journal of Environmental Biology*, 23, 365-71.

Dewettinck, T., Van Houtte, E., Geenens, D., Van Hege, K. and Verstraete, W., 2001. HACCP to Guarantee Safe Water Reuse and Drinking Water Production—A Case Study, Water Science and Technology, 43, 31–38.

Kapley, A. and Purohit, H.J., 2001. Tracking of Phenol Degrading Genotype, Environment Science and Pollution Research, 8, 89-90.

Kapley, A., Lampel, K. and Purohit, H.J., 2001. Rapid Detection of Salmonella in Water Samples by Multiplex PCR Water Environment Research, 73, 461-65.

Kutty, R., Kapley, A. and Purohit, H.J., 2001. Pseudomonas sp. Strain RM2: Strain with Diverse Physiology for Aniline and Chlorophenol Utilization. *Asian Journal of Microbiology Biotechnology and Environment*, 3, 117-21.

Qureshi, A., Prabu, S.K. and Purohit, H.J., 2001. Isolation and Characterization of Pseudomonas Strain for Utilization of 4-nitrophenol, Microbes and Environments, 16, 49-52.

Cavalca, L., Di Gennaro, P., Colombo, M., Andreoni, V., Bernasconi, S., Ronco, I. and Bestetti, G., 2000. Distribution of Catabolic Pathways in some Hydrocarbon-degrading Bacteria from a Subsurface Polluted Soil, Research Microbiology, 151, 877-87.

Kapley, A., Lampel, K. and Purohit, H.J., 2000. Thermocycling Steps and Optimization of Multiplex PCR Biotechnology Letters, 22, 1913-18.

Kutty, R., Purohit, H.J. and Khanna, P., 2000. Isolation and Characterization of a Pseudomonas sp. Strain PH1 Utilizing Meta-aminophenol, *Canadian Journal of Microbiology*, 46, 211-17.

Seckler, D., Barker, R. and Amarasinghe, U., 1999. Water Scarcity in the Twenty-first Century, *International Journal of Water Resources Development*, 15, 29-42.

Talley, J.W. and Sleeper, P.M., 1997. Roadblocks to the Implementation of Biotreatment Strategies, Annual NY Academy of Science, 829, 16-29.

Chhatre, S., Purohit, H., Shanker, R. and Khanna, P., 1996. Bacterial Consortia for Crude oil Spill Remediation, Water Science and Technology, 34, 187-93.

Ka, J., Holben, W. and Tiedje, J., 1994. Use of Gene Probes to Aid in Recovery and Identification of Functionally Dominant 2, 4-dichlorophenoxyacetic Acid-degrading Populations in Soil, Applied Environmental Microbiology, 60, 1116-20.

Van der Meer, J., de Vos, W.M., Harayama, S. and Zehnder, A.J.B., 1992. Molecular Mechanisms of Genetic Adaptation to Xenobiotic Compounds, *Microbiology Reviews*, 56, 677-94.

Hall, E.R. and Melcer, H., 1983. Biotechnology Developments for the Treatment of Toxic and Inhibitory Wastewaters, *Biotechnology Advances*, 1, 59-71.

20

Industry Based Issues (CREP)—Zero Discharge

T. Nandy

Preamble

The concept of zero discharge means in industrial parlance, reduce, reuse, and recycle the effluents generated within the industry through waste minimization and treatment techniques, so that no effluent is discharged from the industry. Zero discharge concept is realized through utilization for agriculture, green belt development, fire fighting, water for washings inside the plant and outside area, process use, cooling water makeup, maintenance of recreation area viz., artificial lakes, impoundments for biodiversity. The benefits of zero discharge are increasingly recognized as an essential strategy in most of the parts of the world where water is in short supply. Wastewater reuse has two major objectives: it improves the environment because it reduces the amount of waste (treated or untreated) discharged into watercourses and it conserves water resources by lowering the demand for fresh water abstraction. In the process, zero discharge has the potential to reduce the cost of both wastewater disposal and the provision of irrigation water, mainly around towns with sewers.

The benefits of zero discharge are several, including:

- Reduction in freshwater extraction for the intended use.
- Enabling the reallocation of freshwater supplies for urban use.
- Beneficial use on application of treated effluent, decreasing the use of fertilizers while providing increase in nutrients and organic matter for soil conditioning.

- Soil conservation and potential reduction of desertification.
- Improvement of the environment by enabling zero-discharge to receiving bodies.

Current Status

Initially the wastewater generated from the industries is treated through conventional primary (physico-chemical) and secondary treatment (biological route) before polishing the effluents at tertiary stage for the intended use. Some of the tertiary treatment options adopted alone or in combination with reuse are presented hereunder:

Filtration

Filtration is a polishing process employed to remove low concentration of suspended solids from wastewater effluents of biological and chemical treatment processes, and is used as a pretreatment method to protect more sensitive adsorption, membrane separation and ion exchange equipments.

The complete filtration operation comprises two phases: filtration and cleaning or backwashing. The wastewater to be filtered is passed through a filter bed consisting of granular material (sand, anthracite and/or garnet), with or without added chemicals. Within the filter bed, suspended solids contained in the wastewater are removed by means of a complex process involving one or more removal mechanisms such as straining, interception, impaction, sedimentation, flocculation and adsorption. The cleaning/backwashing phase follows for cleaning of the media (Chellam *et al.*, 1998).

Disinfection

Disinfection is the process employed for destruction of pathogenic micro-organisms and prevents waterborne diseases. This process is of importance in wastewater treatment owing to the nature of wastewater, which harbours a number of human enteric organisms that are associated with various waterborne diseases. This is particularly important when the secondary effluent is discharge into a water body used for swimming or for a downstream water supply.

Commonly used means of disinfection include the following:

- Physical agents such as heat and light.
- Radiation, mainly gamma rays.
- Chemical agents including chlorine and its compounds, bromine, iodine, ozone and hydrogen peroxide.

The most common chemical disinfectants are oxidizing chemicals, and of these, chlorine is the most widely used. Disinfectants act through one or more of a number of mechanisms, including damaging the cell wall, altering cell permeability, altering the colloidal nature of the protoplasm and inhibiting enzyme activity. In applying disinfecting agents, several factors need to be considered: contact time, concentration and type of chemical agent, intensity and nature of physical agent, temperature, number of organisms, and nature of suspending liquid.

Adsorption

Adsorption is employed to remove a large variety of dissolved organic solutes, and a limited number of inorganic solutes from wastewater contaminants, using adsorbent materials such as activated carbon with a high adsorptive surface area (500-1500 m^2/g). Adsorption of organic solutes is commonly followed by thermal regeneration of the carbon and simultaneous destruction of the adsorbates. In a few cases, the carbon may be regenerated and adsorbate recovered by treatment with acid, base, steam and solvent. In wastewater treatment, adsorption with activated carbon usually follows normal biological treatment, and is aimed at removing a portion of the remaining dissolved organic matter.

A fixed bed carbon adsorption system is typical of industrial wastewater treatment systems. Wastewater flows through a vertical column containing activated carbon that adsorbs organics on a highly porous surface. Adsorption systems are usually designed as continuous processes, configured for cocurrent or countercurrent application. In any adsorption system, performance is measured by contact time and carbon capacity.

Softening

Softening is the process of reduction of hardness in the wastewater for subsequent use. Softening can be achieved by chemical precipitation and ion exchange. In chemical precipitation, pH is adjusted to alkaline range (pH 10.6) by addition of lime and precipitation is accomplished by converting calcium and magnesium hardness to calcium carbonate and magnesium hydroxide, this is normally achieved by lime soda ash process. In ion exchange, calcium and magnesium ions are removed by interchanging with sodium as they pass through a bed of ion exchange resins.

Photochemical Oxidation

Direct photooxidation with ultraviolet light for oxidation of organic pollutants, can be made. However, there are a number of limitations in its use. The first one being that the organic compound to be eliminated must absorb light in competition with other compounds of the effluent to be treated. The second one is that the organic compounds generate a wide variety of photochemical reactions (secondary pollutants) that are more complex to degrade.

The addition of energy as radiation to a chemical compound is the principle of the photochemical processes. The molecules absorb this energy and reach excited state, and chemical reactions are initiated. In the photochemical reactions, hydroxyl radicals may be generated by water photolysis (Cervera and Esplugas, 1983)

$$H_2O \xrightarrow{hv} H^{\bullet} + OH^{\bullet}$$

Photolysis involves the interaction of light with molecules to bring about their dissociation into fragments. This reaction is a poor source of radicals, and in the reaction medium large quantity of reaction intermediates that absorb part of the radiation are generated, which decrease considerably the photooxidation kinetics of the contaminants. That fact makes the process valid only for effluents with low concentration of pollutants.

The photochemical treatment, although partially solves the problem of the refractory compounds, has some limitations in its application, due to higher cost on UV radiation generation (Rodriguez, 2003). Furthermore, not all the emitted radiation is used, only the absorbed radiation produces chemical changes. This fact facilitates some photo degradation reactions with very low yields and slow kinetics. To accelerate the process, other oxidants like hydrogen peroxide and/or ozone, metallic salts or semiconductors like TiO_2 can be added, giving rise to the so-called Advanced Oxidation Processes.

Demineralization (Ion Exchange)

Ion exchange is employed to remove specific dissolved ions from industrial wastewaters to polish treated effluents for producing water quality effluent for industrial reuse. Typically, wastewater flows through a fixed, vertical ion exchange column, where specific anions or cations are removed on highly active resin surfaces. The principle characteristic of ion exchange is that the resin ion selectivity removes specific inorganics. Its primary application is removing cations such as calcium, magnesium, manganese and iron, and anions such as sulphate, nitrate and chloride. Ion exchange entails a chemical reaction in which stoichiometric quantities of solution electrolyte contact and reacts with an equivalent insoluble resin electrolyte.

Ion exchange technology is primarily employed in specialized processes such as water softening, boiler feed treatment and preparation of mineral free water for manufacturing processes. Ion exchange systems can be configured as fixed bed or continuous bed operations. Effective performance of ion exchange depends on resins selectivity. Total influent dissolved solids are limited to around 1000 mg/l, above which selectivity is difficult to predict, effluent quality significantly deteriorates, and regeneration requirements affects the operating cost. Economical ion exchange applications require optimum use of bed capacity and effective regeneration.

Membrane Separation

Membrane separation technology allows for controlled particle size exclusion through the selection of the pore sizes within the membrane. In membrane process, liquid passes through the membrane; whereas the bulk of the solute is concentrated and rejected. The primary advantages of membrane technology over existing separation processes such as distillation are low energy cost, low cost, modular construction and high selectivity of separated materials.

Membrane technology has found increasing application in many industries that require efficient chemical separation, purification, concentration and fractionation. In wastewater treatment, membrane technology is employed to remove specific wastewater components such as dissolved organics and inorganics to produce reusable water or to recover product from specific wastewaters.

The various pressure driven membrane processes include microfiltration, ultrafiltration, nanofiltration and reverse osmosis. The primary difference between the four pressure driven membrane processes is the size of the particles that the membrane rejects, which is determined in part by the pore size of the membrane. The particle size that is rejected by a membrane is given in microns; however, membranes are also classified by molecular weight cut-off (MWCO).

The basis of membrane process is the semi-permeable membrane used for separation, which is the heart of the process. Separation may occur on the surface or within the porous matrix of the membrane. Separation of the mixture occurs, if there is a significant difference in the transport co-efficient through the membrane for the components of the mixture. Membranes can be made of organic polymers, metals, ceramics, and layers of chemicals, liquids and gases. The traditional materials used in the fabrication of membranes include cellulose acetate, polyamide and polysulphone.

The membrane systems are generally designed with one or more modules. The membrane modules currently in practice comprise tubular, spiral, hollow fiber and plate and frame. However, only the first three modules are used on a commercial scale.

Membrane filtration processes are used to remove various dissolved organic compounds from municipal and industrial wastewaters (Mallevialle *et al.*, 1996; Brindle and Stephenson,

1996). The removal of inorganic contaminants by NF and RO remains the largest application in water treatment. In addition, RO and NF have been recently investigated to remove hardness nitrate, and heavy metals (Rautenbacg and Groschl, 1990; Waypa *et al.*, 1997).

According to the CPCB Charter on Corporate Responsibility for Environmental Protection (CREP), the major industries in India must comply with the National Standards notified under the Environment (Protection) Act, 1986 through any or combination of selected treatment technologies and committed to achieve zero discharge (http://www.cpcb.nic.in/Charter/charter5.htm).

One of the major refineries in India, Chennai Petroleum Corporation Limited, Manali, Chennai has installed a state-of-the-art technology for reuse of water from domestic sewage. It employs an effluent treatment plant comprising of primary (sedimentation) and secondary (activated sludge process) treatment facilities to treat the domestic sewage (2.4 MGD from the city). Tertiary treatment of the effluent followed by reverse osmosis process (membrane separation) is carried out. While the state government earns revenue through selling of domestic sewage, the treated effluent from the reverse osmosis caters to about 40% of the plant consumption. Maximum quantity of the water is reused within the refinery complex for fire fighting, green belt development, service water, floor washing, toilet flushing and chemical preparation.

International Concern

One of the major problems faced by the industry worldwide is the identification of safe and economically viable alternative for wastewater reuse with the zero discharge concepts.

Most Serious Issues

Industry's Perception of the Concept

Industry perceives CREP as a regulatory tool to installing expensive treatment system, which increases the cost of production *vis-à-vis* competitive global market. Indian industry is dominated to a large extent by SME's (Small & Medium

Enterprises). Most of these units are located in the unorganized sectors with less disclosure to the authorities. These units must be brought under CREP fold to benefit the objective of the policy.

The Cost of Implementation of Zero Discharge to the Industry (large scale vs small scale)

The main objective of CREP is to introduce industry responsibility for environmental protection by reducing pollutants discharged into environment as well as optimize use of natural raw materials. As a result of accepting CREP, industry has to continuously improve upon its target to meet environmental standards. Development of near zero discharge may be uneconomic for most of the industries at this time, even if the technical know is available. However, the concept will become increasingly attractive as full cost recovery for water and waste disposal are introduced and as technologies for reuse and recycling become more widely available and economically attractive.

Emerging Issues

Merits and Demerits of the Concept and Technologies Available

Even though the concept has not taken wing, increasing demand on water resources and environment may force industries in future to adhere to zero discharge of effluents and emissions. Accepting at the outlet, increases public image of the industry.

Major problem with CREP is that it identifies only the pollutants problem from the industry. It does not delineate or elaborate existing and latest technologies to achieve the norms and costs towards achieving them. Industry does not get any visible benefit out of following CREP. Subsidies or any market-based incentive may enhance prospects of CREP.

Inadequacy of Management and Control

Proper Operation and Maintenance of Treatment Units for Optimum Performance

The major problem faced in meeting the environmental standards by the industry is basically due to operation and maintenance of the existing treatment plants. Addition of new technologies to meet the objective of CREP do not improve the environmental conditions. Skilled and trained personnel are necessary for operation and maintenance of the various unit operations and processes.

Lack of Monitoring

Periodical visits by the regulatory agencies concerned to the industry may improve understanding of the problem by the industry and come out with remedial measures to meet. Major impediments is the shortage of sufficient manpower to make frequent visits to the problem units. Provision of online monitoring and remote data storage system may mitigate the same.

Future Challenges

Water Use Minimization

Globally, water consumption per unit of product across various industries has reduced significantly. Use of latest processing techniques and machinery have reduced water consumption in major industries such as pulp and paper mills, textile units, thermal power plants, tanneries, chemical industries and petroleum refining. Waste minimization is one of the strategies adopted for prevention of industrial pollution. Indian industries need to introduce latest technologies to minimize the water use.

Process Optimization

Process optimization includes use of optimum use of operating parameters (temperature and pressure) and chemicals (quantity of raw material). Non-optimization of process parameters leads to wastage of chemicals, which finally ends up in waste treatment.

Research in the Area of Chemical Recovery from Potential Industrial Effluents

Membrane and evaporation systems offer much promise towards the area of chemical recovery from potential industrial effluents. Manufacturing of membranes for selective separation of chemicals, from the industrial effluents is an emerging area, and is one of the major challenges in chemical recovery.

Economically Viable Technology for Small-scale Sector

Evaluation and assessment of various best available technologies (BAT) in the country based on criteria such as regulatory discharge norms, operation and maintenance costs, and availability of skilled manpower could provide solution to the problems faced by SME's. Selection of techno-economically viable technology needs to be identified for these industries.

REFERENCES

Brindle, K. and Stephenson, T., 1996. The Application of Membrane Bioreactors for Treatment of Wastewater, Biotechnology and Bioengineering, 49.

Cervera, S. and Esplugas, 1983. Obtencion de Hydrogen Mediante Fotolisis del Agua Energia, 9, 103-07.

Chellam, S., Serra, C., and Wiesner, M., 1998. Estimating Costs for Integrated Membrane Systems, AWWA, 1998.

Mallevialle, J., Odendall, P.E. and Wiesner, M.R., 1996. Water Treatment Membrane Processes, McGraw-Hill, New York, 1996.

Rautenbacg, M. and Groschl, A., 1990. Separation Potential of Nano-filtration Membranes, *Desalination*, 77, 73-84.

Rodriguez, M., 2003. Fenton and UV-vis Based Advanced Oxidation Processes in Wastewater Treatment: Degradation, Mineralization and Biodegrability Enchancement, Universities De Barcelona.

Waypa, J.J., Elimelech, M. and Hering, J.G., 1997. Arsenic Removal by RO and NF Membrane, *Journal of American Water Works Association*, 89(10), 102-14.

CPCB. Charter on Corporate Responsibility for Environmental Protection (CREP), Distilleries. (http://www.cpcb.nic.in/Charter/charter5.htm)

Water Renovation

T. Nandy

21

Preamble

Availability of water in adequate quantities and desired quality is a primary requirement for industrial growth, a key parameter to overall development of any region. However, water is becoming a scarce commodity in many parts of the world and these areas have turned to reclaiming wastewater as a way to increase water availability supplies and maintain the pace of development. Wastewater renovation and reuse have emerged as practical supplemental strategies for water resource development and management (Bartone, 1991). The advantages associated with the wastewater, especially municipal wastewater, are its assured year round availability, stable composition, and easy amenability to renovation/treatment. The effluent quality requirement, depending upon the intended use, can also be met through application of one or a combination of the available treatment technologies (Ali, 1987).

Major Issues

Water requirement in India for domestic and industrial purposes are constantly increasing. Some of the important issues that need consideration include (Urban Statistics, 2000) following:

- 200 million people in India do not have safe drinking water and the figure has risen to one billion in 2005 (WHO Report, *Hindu*, 2005).
- 1.5 million children under 5 years of age die annually from waterborne diseases.

- Sewage treatment facilities are inadequate in most of the cities and almost absent in rural areas in the country.
- Domestic sewage, industrial effluents and agricultural runoff pollute more than 70% of water sources.
- Only 60% of large and medium scale industries have appropriate wastewater treatment facilities. Pollution discharges from over 2 million small-scale industries is largely ignored.
- Major portion of the wastewater generated from the cities is discharged without treatment (<6% in Class I cities, <34% in Class II cities is treated) and less than 32% of the wastewater generated from major cities are collected and treated upto secondary level.

Considering the above, renovation of municipal wastewater, to begin with, employing simple and low cost technologies based on land treatment system should be incorporated in any Water Resource Development Management Plan.

Current Status

In India, there are 14 major river basins with natural waters that are being used for human and developmental activities. These activities contribute significantly to the pollution loads of these river basins. The Ganga sustains the largest population, and due to rapid progress in communications and commerce, there has been a swift increase in the urban areas along the river Ganga. As a result, the river is no longer a source of water only, but is also a channel, receiving and transporting urban wastes away from the towns. Today, one third of the country's urban population lives in the towns of the Ganga basin. Out of the 2300 towns in the country, 692 are located in this basin, and out of these, 100 are located along the riverbank itself.

The choice of technology for most of the large municipal wastewater treatment plants was the well accepted activated sludge process, and for other plants trickling filters were considered more appropriate. In smaller towns where land was available and the quantity of wastewater was small, other options such as oxidation ponds were chosen. However, unconventional technologies like the rope bound rotating biological contactors

(RBRC), sewage irrigated afforestation, upflow anaerobic sludge blanket (UASB) technology and plants for chromium recovery from tannery wastewater were tried out with a fair degree of success. An integral part of these sewage treatment works had been self-sufficient from resource recovery by the sale of treated effluent as irrigation water for agriculture, by the sale of dried sludge as manure (because it was rich in nutrients) and from the generation of electricity from the bio-gas production in the plant. It was considered that the generation of bio-electricity would be sufficient to offset much of the cost of the huge energy inputs required.

Industrial wastewater renovation is an emerging area in India with only few industries, which are located in water scarcity area, opting for the concept. Details of such industries are presented in Table 1.

Table 1. Industries with Effluent Recycle Installations

Sl. No.	Industry	Quantity, m^3/hr	Type of Treatment
1.	Arvind Mills Ltd.	10	RO Plant
2.	Ashapura Volclay	20	UF-RO
3.	Chennai Bottling Ltd.	1	RO Plant
4.	Chennai Petroleum Corp. Ltd.	200	UF-RO
5.	Dr. Reddy's Lab	6	RO Plant
6.	Heubach Color	5	UF-RO
7.	Hindustan Lever Ltd.	10	UF-RO
8.	Hyundai Motor India	30	UF-RO
9.	IFFCO	150	RO Plant
10.	INDAL Ltd.	1	Lamella Clarifier
11.	Madras Fertilizers Ltd.	150	UF-RO
12.	Maratha Cements	15	UF-RO
13.	Reliance	2000	UF-RO
14.	Samcor Glass	15	UF-RO-MB
15.	Sundram Fasteners	18	UF-RO
16.	Tecumseh	10	2 Stage RO
17.	Tirupur Arora Textiles	30	RO Plant
18.	TVS Suzuki	12	DM Plant

Source: http://greenbusinesscentre.com/Documents/Designoftreatmentsystems IonExchange.pdf.

The renovated water is recycled or reused within the plant for low end use (toilet flushing and green belt development) as well as high end use (boiler feed, cooling water makeup, process water, etc.).

International Scenario

Globally traditional treatment technologies commonly used to reclaim water include clarification, sand filtration, and disinfection. Wastewater reclamation requires advanced treatment technologies and equipments, such as membrane separation devices, electrodialysis, nanofiltration, ultrafiltration, microfiltration, electrodeionization, ion exchange, activated carbon filter, cartridge filter, ultraviolet disinfection, and ozone generators are also used in this process. Municipal wastewater reclamation demands large-scale treatment technologies, while buildings and residential areas need small-medium scale treatment technologies and equipment.

Wastewater reclamation is a new market strategy that is emerging from China's water scarcity problems. Reused water is priced in some cities and provides a market for wastewater reclamation. Some successful cases of environmental enterprises or companies participating in investing and operating wastewater reclamation projects include a Dalian company, which invested $1.21 millions to construct a wastewater reclamation project with a production capacity of 5300 cubic meters per day, and Qingdao company, which invested $590,000 on similar project. To mitigate water shortages at Beijing, Government constructed centralized wastewater reclamation projects and supported a number of discrete gray water reuse projects. The volume of reclaimed wastewater in Beijing is currently over 147 million cubic meters per year and it is planned to increase to 600 million cubic meters per year by 2008 (U.S. Dept. of Com., 2005).

In general, technology and equipments with high efficiency, low cost, and convenient operation and maintenance have the greatest sales potential in future.

International Concerns

- Current practice of water being used as conveyance media for pollutants from domestic and industrial sector needs to be replaced by other options viz. on site treatment and other new technologies to reduce the cost of treatment through ecological engineering or natural treatment based systems.

- The geo-political situation arising out of increasing scarcity of fresh water is much more difficult to manage as every country wants to meet its public water requirement. Whether or not a third world war takes places, it is certain that the next war would be fought for control of water rather than for oil.
- The public in general is spending more time and money on meeting their water demand. As a result precious man-hours and capital are lost.
- Existing guidelines and standards for wastewater recycle/reuse do not cover some pollutants such as Persistent Organic Pollutants (POP's), pesticides, organo-chlorines, etc.

Most Serious Concerns

- Seggregation of domestic and industrial effluents in order to facilitate resource recovery.
- Identification of industrial effluents and processes where water renovation could be implemented.
- Reduction of water quantity requirements in industries through reuse/recycle techniques and use of cleaner technologies, wherever possible.
- Highest priority for detoxification of elements and compounds (heavy metals).

Emerging Issues

The data on studies on long-term effects of certain components (viz. Carcinogens) in the renovated water on human beings, plants, animals and ecosystem, in general, is presently not available in this country. Therefore, prior to introducing renovated wastewater in the industry, a scientific study on epidemiological, physiological and ecological aspects must be undertaken.

Management and Control

Implementation of suitable technologies so as to sustain operation and maintenance of the treatment plants:

- Training of personnel to meet the challenges related to new technologies for wastewater renovation.
- Creating adequate database on ecological impacts of the use of renovated water in the country with associated health implications on human, plant and animal life.

Future Challenges

- Increased efforts should be made to study the socio-cultural aspects of reusing renovated wastewater, to promote and enhance its acceptance by removing barriers through proper education and interaction with the masses.
- Use of renovated wastewater should be strictly monitored for the intended use.
- National policy on wastewater renovation must be framed in consultation with the Central Pollution Control Board, State Pollution Control Boards and other concerned agencies/organization.

REFERENCES

U.S. Department of Commerce, Washington, D.C., 2005. Report on Water Supply and Wastewater Treatment Market in China.

Urban Statistics, Hand Book, National Institute of Urban Affairs, 2000.

Bartone, 1991. International Perspective on Water Resources Management and Wastewater Reuse Appropriate Technologies. *Water Science Technology*, 23, 2039-47.

Ali, I., 1987. Wastewater Criteria for Irrigation in Arid Regions. *Journal of Irrigation and Drainage Engineering*, 113(2), 173-83.

Indiastat. (http://www.indiastat.com/India/Environment and Pollution/ Water Pollution)

Wastewater for Land Use

22

A.A. Juwarkar

Preamble

Wastewater usage for land application in agriculture started mainly as an alternative to prevent pollution of water bodies, rather than to enhance crop production. The scarcity of water for irrigation warrants that we utilize the knowledge of the beneficial effects of wastewater usage on crops. Currently, India uses 396 billion cu. metres of groundwater (325 for irrigation and 71 for domestic use) (Singhal, 2003). It is estimated that the requirement of water for irrigation by year 2010 will increase to about 556 billion cu. metres. The current utilization and also need for water has witnessed reduction of per capita availability of fresh water from 5,200 cu. metres in 1947 to 1,950 cu. metres in the year 2000 (CWC Report, 2000).

Wastewater can also be utilized for forestland development. The National Forest Policy (1986) envisages that 33% of land needs to be forested. According to Central Statistical Organization, Indian forest cover in the year 2001 was only 20.55% (http://www.envfor.nic.in/nfc/fc-stat.htm, Juwarkar *et al.*, 1990). Crops irrigated with wastewater produce better yield. A long-term field experiment at NEERI has shown that medium intensity irrigation with settled wastewater increases the crop yield by 22.6% on an average than crops irrigated with standard doses of NPK (Duncan Mara, 1987). The total economic value of sewage generated from class-I cities in India, assuming no loss of nutrients after treatment works out to be Rs. 1595.05 million per year (CPCB, 2000). Therefore, wastewater application on land to raise food crops and forestry can help in conserving nutrients available in wastewater and also in meeting the water demand for irrigation, forestry as also groundwater recharge.

Current Status

Wastewater use for agriculture in India came into practice as early as 1895 in Ahmedabad. By the year 1985, there were over 200 wastewater irrigation schemes in operation, covering approximately 73,000 ha of land (Duncan Mara, 1987). However, excessive wastewater application rates were used (up to 12 m/year) leading to very high nutrient loading rates (up to 600 kg total N per ha per year) and consequent high built up in the soil (Duncan Mara, 1987). The major wastewater irrigation schemes are operational in Ahmedabad, Amritsar, Bikaner, Bhilai, Delhi, Gwalior, Hyderabad, Jamshedpur, Kanpur, Chennai, Madurai, Trivandrum and Lucknow. At Nagpur, both domestic and Industrial wastewaters (pulp and paper, distillery wastewater) have been investigated to assess feasibility of land application (Juwarkar *et al.*, 1990; NEERI Report, 1998).

In India, 12 metropolitan cities, 212 Class I and 241 Class II towns generate 2358, 4433 and 467 million cu. metres of wastewater per year. Due to huge energy and costs involved in treatment processes, only 26% wastewater in metropolitan cities, 22% in Class I cities and 5% in Class II towns is treated mostly to primary levels (CPCB, 2000). Alternative methods of wastewater treatment include wetlands and high rate transpiration system (HRTS). The Kaladeo National Park at Bharatpur, wetlands of Kolkata, flood plains of Ganges (843,221 ha) and Brahmaputra are natural wetlands used for wastewater treatment. Constructed Wetlands are of two types: Free water-surface and Subsurface flow system. Construction of wetlands involves digging trenches and filling it with selected media and planting with vegetation. They have shown good efficiency by reducing BOD and SS by 77% (Trivedy, 1998). In an HRTS, soil system is used as a living filter to renovate wastewater by adsorption, ion exchange, precipitation and stabilization of pollutant through microbial degradation. HRTS enables wastewater disposal of 250 to 450 m^3/ha/day. HRTS has been applied by NEERI at Amlai on 80 hectares of land and wastewater is applied at the rate of 150-250 m^3/ha/day (NEERI Report, 1998; Singhal, 2003).

At present, treatment of distillery effluent for safe disposal on land is too expensive. The Central Pollution Control Board evaluated the technology for disposal of Distillery wastewater on land in 1998-99 and suggested that the wastewater should undergo partial treatment by way of bio-methanation, followed by aerobic secondary treatment (activated sludge process) and dilution (CPCB, 2000-2001). Indian Agriculture Research Institute has monitored the ferti-irrigation of agricultural crops with distillery effluent at Daraula Sugar Mills in Uttar Pradesh recently. NEERI had undertaken feasibility studies to use treated distillery wastewater for forestry and has shown that distillery wastewater can be used for forestry after proper amendments (NEERI Report, 2003).

International Concern

Internationally, in many countries, strict rules are in place for the application of wastewater for land irrigation, however, the same guidelines and implementation are lacking in many countries, including India. In the current scenario of food products being transported to various countries, it is essential to know the risk of such practices on the produce, particularly food products. Therefore, there is a need to conduct studies to evaluate risks linked with application of wastewater, and in particular, the risks linked to the consumption of irrigated garden products. While wastewater irrigation schemes have a vast scope in developing countries like India, they lack sufficient infrastructure to manage a wastewater land application (WLA) system. As a result, there is a danger of mass health hazards to consumers and farm workers.

Detailed epidemiological surveys of the transmission of enteric viruses via wastewater irrigated land; its products and seafood are also required. Toxic metals may enter food chain through crop contamination. Germination and pot-culture studies at NEERI have shown that germination of gram and groundnut was badly affected due to metallic contamination whereas no adverse affect was observed on wheat, barley, maize, jowar, paddy, mung and urad crops.

Inadequacy in Management and Control

The principle of "to irrigate only when it is required and only to the extent required by the crop" should be the guiding principle in a WLA system. Water requirement depends upon soil type, crop and climate. Over irrigation may result in crop wilting and subsequent damages. A wastewater land application (WLA) system needs mainly two types of monitoring:

- Site Management/Process Control Monitoring.
- Compliance monitoring.

Site Management / Process Control Monitoring: This includes effectiveness monitoring of best application practices, treatment, performance and monitoring of soil, soil-water and plant status.

Compliance Monitoring: It involves determination of whether a WLA system is complying with primary and secondary water quality standards.

An environmental monitoring programme should have monitoring of groundwater, soil water, soil, subsurface water level, wastewater and plant tissues.

Groundwater Monitoring: Monitoring wells defining the hydro geological system, to study the effect on groundwater have to be installed. Periodic observations of four parameters will normally be sufficient for assessment of detrimental effects of land application of wastewater (NEERI Report, 1979) (a) DO, (b) EC (for TDS), (c) H-ion conc. (pH), (d) Turbidity (undissolved solids).

Wastewater Monitoring: The wastewater applied has to be studied normally for parameters like pH, EC, TDS, permeability, phytotoxicity, Sodium, Chlorine, Boron and Nitrogen.

Plant Tissue Studies: Plant tissue parameters like protein concentration, nitrogen, ash and moisture content should be monitored in the WLA site vicinity.

Soil Monitoring: The frequency of soil parameters monitoring will depend on the quality of wastewater applied. Normally, nutrient monitoring and soil quality monitoring is carried out. In nutrient monitoring, parameters such as N, P, K, pH, micronutrients, etc. are monitored. Whereas, in soil quality monitoring, EC, SAR, organic matter, concentration of phytotoxic constituents (like Cl) salinity, concentration of redox-sensitive species (iron and manganese) are to be monitored (Idaho Govt., 2004).

Most Serious Concerns

Farm Workers Health: A study on prevalence of hookworm and ascaris infections in sewage farm workers found that they were 40% more susceptible to infections as compared to non-sewage workers (Duncan Mara, 1987). The enteric pathogens like Coliforms, E.Coli, F.Streptococci survive in clay, loam and clayey soils applied with wastewater that is untreated, 1:1 diluted and primarily treated. These pathogens survive for 120-150 days in sufficient numbers, so as to be harmful to humans (Juwarkar *et al.*, 1998).

Excess Hydraulic Loading: The existing wastewater disposal farms receive 76% of untreated and 24% of primary or secondary treated wastewater (CPCB, 2004). The hydraulic loading rate is different for each site. Normally the average application rates vary from 45 to 55 cu.m/ha/day (Juwarkar *et al.*, 1998). However, the application rates in India vary from 22 to 336.8 cu.m/ha/day. This clearly indicates that wastewater application rates are higher and problems like alkalinity, salinity and soil sickness are common.

Excess Nutrient Loading: The tendency of over irrigation also results in excessive input of plant nutrients. The Nitrogen loading through wastewater varies from 403 to 6052 kg/ha/annum, which is 2 to 17 times the normal crop need. The excess Nitrogen applied in excess of assimilative capacity results in ammonia volatilization, denitrification and leaching of nitrate to sub-soil and to groundwater. Excess Nitrogen (as nitrate or nitrite) in groundwater due to leaching leads to diseases like Methemoglobinemia (blue baby syndrome) or diuresis.

Emerging Issues

Changes in Soil Characteristics: Soils irrigated with wastewater revealed steady build-up of salinity and alkalinity with wastewater application. In a two-year pot-culture evaluation for soils irrigated with wastewater at NEERI (1980), exchangeable sodium percent (ESP) increased by 58.5% and organic matter increased by 9%. But by the end of 6 seasons irrigation (3 years), soils became alkaline and needed reclamation (NEERI Report, 1979).

Toxic Effects: Elements like Boron, chloride, nitrates, etc., which accumulate in soils over a period of time, have undesirable affects on plants and vegetables.

Organic Solids: Excess application of unstable organic matter leads to Oxygen depletion in soil. Deposition of sediments containing clay and colloidal matter may lead to crust formation, impeding seedling development and reducing infiltration.

Some of the above possible changes which have been reported need to be assessed on a continuous basis to understand the possible changes and take appropriate precautions at every stage. The learning curve in this field needs to be documented and shared for better management.

Future Challenges

In India, currently, most of the WLA systems are disposal rather than utilization oriented. The WLA system is not efficiently managed due to application of high hydraulic loading and direct use of untreated wastewater. It is, therefore, essential that the existing system is upgraded with development of design protocol of WLA system, define guidelines of design and delineate process control monitoring system for effective management of WLA system. It is, therefore, a challenge to set guidelines for design and effective management of a WLA system.

Research is needed to gather information on the micro level for phenomenon such as ratio of soluble metal to total soil metal content, specific crop tolerance limits, metal uptake mechanisms and long-term behaviour of soil. Future of land application of wastewater would entirely depend upon the development of process outline and its constant surveillance for any remedial measures.

REFERENCES

Idaho Govt., 2004. Department of Environmental Quality, Idaho Govt., Guidance for Land Application of Municipal and Industrial

Wastewater, Web-based Revision, 2004. (http://www.deq.idaho.gov/water/permits_forms/permitting/guidance_wlap.pdf)

NEERI, 2003. Feasibility Studies on Utilization of Biologically Treated Distillery Effluent for Forestry, sponsored by M/s Jubilant Organosys Ltd., Noida.

Vikas Singhal, 2003. Indian Agriculture, Indian Economic Data Research Center, No. 48, 59.

CPCB, 2000. CPCB Manual on Status of Water Supply and Wastewater Generation, Collection, Treatment and Disposal in Class I cities by Central Pollution Control Board, Feb. 2000, No. 20, 24.

CPCB, 2000. (www.cpcb.nic.in/annual_report 2000-01-34.htm)

NEERI, 1998. Treatment and Disposal of Wastewater through High Rate Transpiration System at Orient Paper Mills, Amlai.

Juwarkar, A.S., Juwarkar, A., Deshbhratar, P.B. and Bal, A.S., 1990. Exploitation of Nutritional Potential of Sewage and Sludge through Land Application, Expert Consultant Meeting held at Food and Agriculture Organization, Bangkok.

Duncan Mara and Sandy Caincross, 1987. Wastewater and Excreta use in Agriculture and Aquaculture in Developing Countries, United Nations Environmental Programme and World Health Organization, 1-3, 1-23.

NEERI, 1979. Utilization of Pulp and Paper Mill Wastewater for Agriculture sponsored by Orient Paper and Industries Ltd., Amali.

Juwarkar, A.S., Juwarkar, A.A., Thawale, P.R. and Jambulkar, H.P., 1998. Management of Wastewater through Crop Irrigation—An Eco-friendly Approach, Eco-technology for Pollution Control and Environment Management, 25-48.

Trivedy, R.K., 1998. Use of Constructed Wetlands in Wastewater Treatment: The Root Zone Technology, Ecotechnology for Pollution and Environmental Management, Enviro Media, Karad, India, 223-28.

MOEF, 2001. State of Forest Report, 2001. (http://www.envfor.nic.in/nfc/fc-stat.htm)

SOLID WASTE

23

Inventorization and Characterization of Municipal Solid Waste

V.P. Deshpande

Preamble

Municipal solid waste management (MSWM) involves collection, transportation, processing and disposal of solid wastes (SWs) generated in a city. The continuously increasing size of the urban population makes it progressively more complex and difficult to manage the waste (Tchobanoglous *et al.*, 1993). To analyze the activities associated with SWM from the point of generation to final disposal, these can be grouped into six elements i.e. waste generation, on-site handling, storage and processing, collection, transfer and transport, processing and recovery and disposal. Since the system handles huge quantity of waste, it becomes necessary to have detailed information on sources of SW generation, their quantification and characterization so as to plan different elements for handling the waste. Presently in India, the MSWM system is not planned rationally due to non-availability of authentic or relevant information on waste generation and its characterization (Kumar, 2005).

Quantification

The quantity of MSW reaching the disposal site is arbitrarily estimated based on number of trips performed by the transportation vehicles. Further, the quantity that is indicated often represents the quantity that has been removed and not the generated quantity. Hence, the collection system should be source specific (NEERI, 1996).

Sampling

The MSW samples are collected from selected points based on population density, number and location of market places, commercial and business areas scattered throughout the city. The waste characteristics from various categories are averaged, given a weighing factor, and added to obtain MSW characteristics for the entire city. The weighing factor for a given category is the ratio of the weight of MSW produced by the population in a given category to the total waste produced by the city. For sampling usually cornering and quartering method is adopted. In this method, large quantity of waste is well mixed to make the characteristics more uniform and a pile (coning) is formed wherefrom a portion is randomly selected. This procedure is performed number of times to obtain single sample for further analysis (CPHERI, 1973).

Sampling technique adopted by Warren Spring Laboratory in UK involves separation of waste by size using a rotating drum with 160 mm dia (aperture) and designed specially for the purpose of screening raw waste. The waste is seggregated into two streams viz. oversize (+160 mm dia and undersized –160 mm dia). Samples of the oversize and undersize streams are taken at 5 to 10 minutes intervals beginning approximately 10 minutes after start up of the trommel. Composite samples of 100 kg weight for oversize and 200 kg for undersize are prepared by adding the incremental samples (EPA, 1996).

ASTM method of determination of unprocessed MSW describes procedures for measuring the composition by manual sorting. The method gives mean composition of MSW based on the collection and manual sorting of a number of samples of wastes over a selected time period covering a minimum of one week. As per this method, vehicles for sampling are selected randomly during one-week sampling period. About 1816 kg of sample is reduced to 454 kg by quartering technique. This method is suitable for house-to-house collection system adopted in western countries (www.webstore.ansi.org/ansidocstore/product.asp?sku=ASTM+D5231% 2D92%282003%29).

Physical Characteristics

The representative samples are sorted into separate sections such as stone, coal, vegetable matter, glass, paper, rags, wood, iron, fine earth and others and weighed. These fractions are then represented as percent by weight. From this, the total percentage of compostable materials is calculated. After physical sorting, fractions like metal, plastics and rubber, etc. are segregated, and set aside, moisture content is determined from the remaining crude samples (CPHERI, 1973).

Chemical Characteristics

After the moisture is estimated, a sample of about 3 kg is air-dried, finely grounded, and passed through a fine sieve of 0.0166-in pore size (CPHERI, 1973). The sample (which may be either raw or composted material) is then analysed for the following parameters:

- Moisture Content
- Organic Matter
- Carbon
- Nitrogen (Kjeldahl)
- Phosphorous as P_2O_5
- Potassium as K_2O
- pH

Extensive studies in more than 40 cities carried out by NEERI indicate that the organic content in the MSW ranges from 30 to 40% and the inert matter has 40-55%. The paper content varies from 3-6%, which is recycled on priority followed by plastic, glass and metals (1%). The C : N ratio varies normally between 20 and 30 and the lower calorific value ranges between 800-1000 kcal/kg. The density varies from 330-560 kg/m^3 due to inclusion of inert matter, drainage silt, demolition waste, etc. (NEERI, 1996).

Municipal Solid Wastes (Management and Handling) Rules, 2000

The Ministry of Environment & Forests (MoEF), Government of India, New Delhi promulgated MSW (Management & Handling) Rules, 2000 with a view to regulate the management and handling of the MSW, which makes it mandatory for every municipal authority responsible for collection, segregation, storage, transportation, processing and disposal of MSW. With the enforcement of these rules, it has become necessary to adopt environment friendly SWM system. Various measures for improvement in existing SWM system have been suggested and incorporated in the notification under schedule I, II, III and IV (MoEF, 2000). The notification necessitates submission of annual performance report on the status of MSWM to designated authorities in each State in India.

Schedule I involves compliance with respect to the setting up of waste processing and disposal facilities, monitoring the performance of waste processing and disposal facilities, improvements at existing landfill site and identification of new landfill sites for future use.

Schedule II deals with the management of MSW with reference to collection of MSW. The provisions include segregation of waste at source, creating storage facilities devoid of unhygienic and insanitary conditions, organizing house-to-house collection of MSW, adopting biological processing of biodegradable waste and non-inclusion of biomedical and industrial wastes in the MSW.

Schedule III relates to specifications for landfill sites with reference to site selection based on environmental issues and location, providing facilities at the landfill site, environmental monitoring, mainly air and groundwater monitoring, plantation to be carried out at and around the landfill site and post-closure plan.

Schedule IV comprises standards for composting, treated leachates and incineration. The waste processing or disposal facilities should include composting, incineration, pelletization, energy recovery or other facility based on state-of-the-art technology duly approved by the Central Pollution Control Board (CPCB).

A recent study carried out by NEERI through CPCB sponsored project on Status of Municipal Solid Waste Management in Metro and Class I Cities reveals that the MSW Rules are not complied with in majority of cities. The major observations are non-adoption of bioprocessing of wastes and simple open dumping of waste. Also, the cover is not provided over the dumped area (NEERI, 2005). Due to scarcity of funds, municipalities are unable to adopt sanitary landfill facility. In view of this, there is a need to think about development of regional landfill site(s).

Status of Solid Waste Management

Solid waste management includes management of various activities like collection and storage, transportation, treatment and disposal. These activities are interlinked and overall efficiency of the system depends on the efficiency of individual activities.

Collection

Waste generation rate in major cities was found to vary between 0.17-0.67 kg/capita/day. In India, mainly community bin collection system is used. These are installed at various locations in the city. Wastes from various sources like residential areas, commercial areas and market places are stored in the dust bin. Road sweepings are collected in the bins. In most of the major cities, old RCC/masonry dust bins are being replaced by 4.5 cu.m. MS container. Improper location and poorly designed dust bins result in low collection efficiency. After the introduction of MSW (Management & Handling) Rules, house to house collection has been initiated in 70% of major cities. However, the activity is mostly limited to identified areas in the cities.

Transportation

Vehicles used for transportation of wastes from collection point to the disposal site include trucks, tipper trucks, dumper placers, tractor and trailer, etc. In many cities, vehicles have outlived their normal life. Poorly designed vehicles and unplanned transportation routes are also the major concern in some cities. In 73% of the major cities, proper covers are not put on the vehicles. This has resulted in odour and spillage problem.

Treatment

Composting of MSW is practised only in 32% of the major cities. However, process adopted is not always environment friendly. Moreover, lack of good quality compost and proper market are also major problems in composting. Refuse derived fuel based power plants are in operation in two cities (Hyderabad and Vijaywada). One biomethanation plant at Lucknow is facing problem of underutilization of the capacity.

Disposal

Sanitary landfilling is rarely practised in India. Uncontrolled dumping in low lying areas is the usual practice of waste disposal. Many municipal bodies have taken initiative for improving the situation as per the directives/guidelines of MSW rules.

REFERENCES

Kumar, S., 2005. Models for Municipal Solid Waste Management, Seminar Report, I.I.T. Bombay, 1-4.

EPA, 1996. Methodology for Conducting Composition Study for Discarded Solid Waste. D.R. Reinhart and Pamela McCauley-Bell (Eds). (http://www.floridacenter.org/publications/discarded_waste_compos_96_1.pdf)

NEERI, 1996. Strategy Paper on Solid Waste Management in India, MUD and PA, New Delhi.

Municipal Solid Waste (Management and Handling), Rules, Sept. 2000, Gazette of India, Extraordinary Reg. No. D.L. 33004/99.

Tchobanoglous, G., Theisen, H. and Vigil, S., 1993. *Integrated Solid Waste Management*, McGraw Hill, 417-29.

Robert, M., Clark, M. and Gillean, J., 1974. Systems Analysis and Solid Waste Planning, *Journal of Environmental Engineering*, ASCE. 7-24.

CPHERI, 1973. Solid Wastes in India, Final Report Prepared Under Foreign Research Agreement, 01-504-01, 10-31.

ANSI. American Nationals Standards Institute document. (http://www.webstore.ansi.org/ansidocstore/product.asp?sku=ASTM+D5231%2D92%282003%29)

Treatment and Disposal of Municipal Solid Waste

24

J.K. Bhattacharya

Preamble

Quantity of municipal solid waste (MSW) is ever-increasing due to various factors like increase in population, change in life style, socio-economic conditions, etc. A number of processing technologies are available for MSW management and hence, adoption of appropriate treatment/disposal system of waste compatible to Indian conditions is needed so that degradation of environmental problem is minimized. In the year 2002, the Ministry of Environment and Forests (MoEF), New Delhi promulgated MSW (Management & Handling) Rules to control the safe movement of MSW right from the point of generation to disposal. The rules also necessitate adoption of environment friendly treatment and disposal system (MoEF, 2000).

Treatment/Processing

Several Processing/Treatment methods are underway to improve better management of MSW. Few important processes are discussed:

Thermal Treatment Process

Thermal treatment processes include incineration, pyrolysis, gasification, etc. Incineration is basically a complete combustion process. High moisture and ash content and low calorific value (800-1000 Kcal/Kg) are the major constraints for incineration of Indian MSW. Moreover, furan and dioxin emissions are also a major challenge. Pyrolysis and gasification are basically incomplete combustion processes. The end products obtained are solid, gas and liquid fuel depending upon the temperature and the process adopted. These processes have been tried for various types of biomass, scrap type, plastic waste, etc. However, technologies for large-scale application are yet to be established in India. Conversion of MSW to refuse derived fuel (RDF) for power generation is also one of the root causes for utilization of MSW. Various steps involved in the process are drying, removal of sands and stones, segregation of combustible fraction and preparation of pellets. Rapid burning is the major drawback of RDF. To overcome this problem, it is often mixed with coal dust.

Biological Treatment Process

Biological treatment processes applied for MSW are composting, biomethanation, etc. Composting is a biological decomposition process of organic fraction of waste to produce humus like materials, which can be applied as a soil conditioner.

India is basically agriculture based country. Hence, composting is a useful option. However, with respect to utilization of the product there is a communication gap, which needs to be filled, so that the product can reach the farmers more easily and composting can become popular. There is also an urgent need to improve the process of composting so that time is reduced and quality is improved. Vermicomposting is a recent advancement in this direction. Efforts have been initiated to convert organic rich waste to compost at small-scale level with the help of earthworms. However, feasibility of large-scale application is yet to be established. Population breeding of earthworms and temperature control are the major challenges.

Biomethanation involves biological conversion of organic fraction of waste under anaerobic conditions. The process has a bright prospect and can provide solution to energy crisis (Ambulkar and Shekdar, 2004). Some of the European countries have developed anaerobic digestion process viz. DRANCO process (Belgium), VALORGA (France/Netherlands). In the Dranco process, mixing occurs via re-circulation of the wastes extracted at the bottom end, mixing with fresh wastes (one part fresh wastes for six parts digested wastes) and pumping to the top of the reactor. This sample design has been shown effective for the treatment of wastes ranging from 20 to 50% total solids (http://www.ees.adelaide.edu.au/pharris/biogas/pvdv.pdf). The processes need size reduction before being treated in the digester. The Valorga digestion process is a semi-continuous, high solid, one-step, plug flow type process. The main characteristic is the complete absence of any mechanical equipment inside the reactions. The waste to be treated consists of food and garden waste (http://www.iwaponline.com/wst/03606/wst036060457.htm).

A biogas generation plant has been reported to be installed at Pune, which utilizes vegetable market waste. 80 m^3 of biogas is generated per day per tonne of vegetable waste. Indian Institute of Science, Bangalore has also attempted biomethanation of domestic Solid Waste. NEERI is in the process of development of two-phase biomethanation process compatible with Indian MSW, where pretreatment of MSW is not required.

Landfilling Practices

In India, usually uncontrolled disposal practice is widely adopted. MSW is usually dumped in low lying area without taking adequate protective measures. In the rainy season, leachate finds access through soil strata and pollutes the groundwater. Surface run off also creates ecological imbalances due to pollution of surface water bodies. Sanitary landfilling involves application of engineering principles for disposal of MSW so that the waste is confined to a minimum volume and soil cover is applied over the compacted waste. Utmost precaution is taken to protect the groundwater from pollution. However, in India, it is rarely adopted.

Groundwater contamination and uncontrolled methane emissions are also one of the major concerns requiring appropriate land disposal. Thus, it is necessary to select appropriate site for developing sanitary landfill system. Criteria for selection of landfill site include haul distance, local restrictions, soil characteristics and topography, climatic conditions, geological and hydro-geological conditions, etc. Use of composite liner of HDPE and clay in lining system and geotextile sheet as filter medium are the recent advances in design of sanitary landfill. Moreover, capture of landfill gas and subsequent flaring up/utilization and leachate collection and treatment need special attention. NEERI has devised criteria for selection of Landfill site (NEERI, 2002). Moreover, it is also necessary to adopt various recycling/treatment technologies to decrease the load on landfills. Source segregation of biodegradable and other recyclables like plastics, papers, glass, metals, etc. will also reduce the load on landfills.

REFERENCES

Ambulkar, A.R. and Shekdar, A.V., 2004. Prospects of Biomethanation Technology in the Indian Context: A Pragmatic Approach. *Resources, Conservation and Recycling*, 40(2), 111-28.

NEERI, 2002. Guidelines for the Selection of Site for Sanitary Landfilling, CPCB, New Delhi.

Municipal Solid Waste (Management and Handling), Rules, Sept. 2000, Gazette of India, Extraordinary Reg. No. D.L. 33004/99.

Vandevivere, P., Baere, L. De and Verstraete, W. Types of Anaerobic Digesters for Solid Wastes, School of Earth & Environmental Sciences. (http://www.ees.adelaide.edu.au/pharris/biogas/pvdv.pdf)

Helene Fruteau de Laclos, Serge Desbois and Claude Saint-Joly, 1997. *Anaerobic Digestion of Municipal Solid Organic Waste: Valorga Full-scale Plant in Tilburg*, The Netherlands, IWA Publishing. (http://www.iwaponline.com/wst/03606/wst036060457.htm)

Special Treatment Methods for Municipal Solid Waste

25

V.P. Deshpande

Vermicomposting

Composting (aerobic and anaerobic) application for converting livestock waste has been traditionally practiced in India. Organic wastes including Municipal Solid Waste (MSW) need to be managed in environmentally sustainable manner. Vermicomposting is the special biotreatment approach.

Vermicomposting is the breakdown of organic material that, in contrast to composting, involves the joint action of earthworms and microorganisms and does not involve generation of heat as in the case of composting. The worms consume organic wastes such as food waste, animal waste and sewage sludge and convert the waste to produce soil conditioner. Advantage of vermiconversion is that particle size reduction takes place due to muscular activity of gizzard, an organ of alimentary canal. The end product is granular in nature and rich in soluble nutrients with relatively high level of humification (stabilisation). However, earthworm activity has limitations due to various factors such as temperature (exceeding 30°C is lethal), toxic effect of heavy metals and initial decomposition products like ammonia gas, organic acids, salinity, etc.

On global level, application of vermiconversion is mostly for livestock waste (solid and liquid), sewage sludge, food waste and fruit processing waste. In India, research on vermicomposting is being given priority in many institutions and organisations. Additionally, many non-governmental organisations and individuals are also contributing, though on a small scale and at domestic level. In the long run, certain issues such as emission of greenhouse gases especially NOx (www.wormresearchcentre.co.uk) and survival of pathogenic microbes in processing waste through vermiconversion are going to be critical.

The most important concern for vermiconversion of municipal solid waste on large scale is that the waste requires pre-processing and preparation before its operation as a batch process. Conversion efficiency is normally much less (around 10%) and needs improvement in the system. Large scale application is also hindered due to management and control of the process in terms of environmental factors like temperature, humidity, etc.

Large scale composting system developed in U.K. considered best condition parameters such as temperature (15°-20°C), moisture content (70-90%), earthworm density (kg/sqm of bed surface) and precomposted waste. The system was operated at sub-optimum condition due to management and control of the process. Huntoon (2000) described four types of systems for medium to large scale vermicomposting viz., windrows, beds, bins and reactors. Feasibility of vermiconversion of variety of organic wastes has been discussed with specific reference to role of plant growth regulators (Sharma and Vasudevan, 1999).

Heavy metals present in the MSW and their effect on vermicomposting was studied by NEERI (1999-2002). The study indicated that cadmium concentration upto 100 mg/kg of MSW is lethal. Pre-decomposition of waste prior to vermicomposting has been reported by Shanti *et al.* (1993).

Thermal Treatment

Municipal solid waste, containing 30-50% organics, has a large potential for processing through thermal route. Various approaches such as incineration, refuse derived fuel, pyrolysis, etc. are being adopted for thermal processing in many developed nations. A major limitation of thermal treatment is that burning of waste transforms hazardous characteristics of waste from solid form to air, water and ash. In view of the perceptions about thermal based technologies as contributors of persistent organic pollutants (POP), alternative approaches are under consideration.

Moreover, a major constraint of thermal processing is that the wastes do not have self-sustained burning due to low calorific value of MSW which is in the range of 800-1000 Kcal/kg of MSW and requires large amount of auxiliary fuel.

Incineration

In developed nations, presorting and preprocessing at landfill is done before the waste is prepared for incineration. The characteristics of waste are such that it has higher percentage of combustible matter i.e. paper, cardboards (30-40%) having higher calorific value.

In India, MSW has 30-40% organic matter and a moisture content of 30-50%. Large proportion of paper, cardboard, etc. is normally recycled. Due to higher moisture content of the MSW, heat value of the waste is greatly reduced and self-sustaining combustion cannot be achieved. For mixed solid waste, the capacity of the incinerator plant required will be large thereby resulting in increase in capital and recurring cost. Preprocessing and presorting of waste is, hence necessary.

Pyrolysis

Pyrolysis involves irreversible chemical change brought about by the action of heat in an atmosphere devoid of oxygen. Synonymous terms are thermal decomposition, destructive distillation and carbonization. In partial combustion, oxygen is present in insufficient quantities to cause complete combustion.

With slow heating at moderate temperature, the products are primarily a non-combustible gas and non-reactive char whereas at high temperature, complete destruction of the molecule is likely to take place and the system would yield more liquid of complete chemical composition. Pyrolysis is carried out at a temperature ranging between 500 and 1000°C.

The char, liquid and gas have a large calorific value. Part of the heat obtained by combustion of either char or gas is often used as process heat for the endothermic pyrolysis reaction. Though a number of lab and pilot investigations have been made, only a few have led to full scale plants.

Presorting of MSW for organic matter is essential before it is subjected to pyrolysis process. This will help in reducing size of the reactor system. In order to maintain temperature of waste mass in the range of 500-800°C in the pyrolysis reactor, heating is done using electrical energy with proper insulation maintained. Thus, the process is costly in terms of money required per tonne of waste. The products obtained would be liquid, char and slag along with gaseous mixture having higher calorific value. Hence, lab scale/demonstration studies are needed for achieving optimum conditions along with the characteristics of fed material, products and cleaning of hot gases before its utilisation.

The issues involved in pyrolysis make the process cumbersome. It is not economical to transport the gas produced over long distances and so the power generation equipment needs to be sited with the pyrolysis/gassification facilities. The process becomes economically viable only when the facilities are located near urban centers. Scrubbing of producer gas at high temperatures is yet to be developed on a large scale plant based on pyrolysis and gasifier.

One of the manufacturing companies has proposed incinerator technology with the name of Cement Lock Technology. It is claimed that this technology destroys all POPs and can be used for all types of waste. The technology has been approved by USEPA.

Plasma Arc Technology/Plasma Pyrolysis Vitrification (PPV)

To overcome the problems arising due to adoption of thermal processing, plasma gasification technology is emerging (which claims to be environment friendly). It is a non-incineration thermal process which uses extremely high temperatures in arc oxygen starved environment. The end products are combustible gas and inert slag. This technology has been developed, perfected and patented by RCLi (USPS technology partner).

This technology for energy/resources recovery from organic wastes, basically uses a Plasma Reactor which houses one or more Plasma Arc Torches. By application of high voltage between two electrodes, these torches generate a high voltage discharge and consequently an extremely high temperature environment (between 5000-14000°C). The hot plasma zone dissociates the molecules in any organic material into the individual elemental atoms while all the inorganic materials are simultaneously melted into a molten lava. The gas output after scrubbing mainly comprises of CO and H_2. The liquefied product is mainly methanol.

The entire process is claimed to safely treat all types of hazardous or non-hazardous materials. It has the advantage of zero NOx and SOx gaseous emissions.

Refuse Derived Fuel

Refuse-derived fuel (RDF) typically consists of pelletized or fluff MSW, which is the by-product of recovery operation. Processing of waste removes ferrous materials, glass, grit, and other materials that are non-combustible. The remaining material is used in RDF. The RDF can be used in dedicated RDF boilers, co-firing of RDF with coal and dedicated RDF fluidized-bed boiler.

As the solid waste from developed countries contains a large paper fraction, it was felt that it could be used as a good fuel. In 1972, a plant was set up in St. Louis, USA wherein a solid waste quantity of 295 tonnes was processed in every 8 hours. In the plant, the incoming solid waste was first subjected to size reduction after which the magnetic metals were removed by a magnetic separator. The remaining material was then passed through a vertical flow air separator where paper got removed. The material was further subjected to size reduction and burnt in a boiler of a 125 MW plant. Due to the encouraging performance of this plant, RDF is being used at a number of other power plants. Thus, the problems faced in using refuse alone for incineration with power generation are avoided.

In India, M/s Shriram Energy Systems Ltd., have set up a waste to energy plant of 6 MW capacity at Vijaywada in Andhra Padesh. The plant uses 500 tonnes per day of processed MSW from Guntur and Vijaywada cities. The pelletization technology has been obtained from TIFAC, Department of Science & Technology, Govt. of India. The power plant was commissioned in December 2003.

Another company M/s SELCO International Ltd., Hyderabad have set up first commercial garbage processing plant in India at Gandhamguda village near Hyderabad in the year 1999. The RDF based power plant (6.6 MW) was commissioned in November 2003. During the first year of operation, the power plant produced 28,391 MW power by processing 95,870 tonnes of MSW.

REFERENCES

SELCO, 2004. Power Generation from Municipal Solid Waste, Presented by SELCO International Limited, Hyderabad on 8-12-2004 at MNES, New Delhi.

Shriram Energy Systems Ltd., 2004. Converting Municipal Solid Waste to Energy, Presented by Shriram Energy Systems Ltd., Hyderabad at MNES, New Delhi on 8-12-2004.

NEERI, 2002. Studies on the Influence of Heavy Metals on the Activity of Earthworms and Vermicompost, Sponsored by Dept. of Biotechnology, Project Sanction No. BT/PRO 909/BCE/12/051/98.

Rhonda Sherman-Huntoon, 2000. Latest Development in Mid-to-large Scale Vermicomposting (From Biocycle Magazine).

Manual on Municipal Solid Waste Management, 2000. Published by Central Public Health and Environmental Engineering Organization, Ministry of Urban Development, Government of India.

Satyawati Sharma and Pootma Vasudevan, 1999. Utilization of Organic Waste through Vermicomposting Route for Plants Growth, the R'99 Congress.

NEERI, 1996. Strategy Paper on Solid Waste Management in India—A Report Submitted to Ministry of Urban Affairs and Employment, Government of India.

Santhi, N.R., Bhoyar, R.V. and Bhide, A.D., 1993. Vermicomposting of Vegetable Waste, Compost Science and Utilization, 1(4), 27-31.

EMRC, Vermicomposting. (http://www.emrc.org.au/glossary_printable.asp)

Methane Emissions from MSW Landfills

26

V.P. Deshpande

Preamble

Indian urban centers are rapidly growing resulting in increase in municipal solid waste (MSW) generation and scarcity of land for the disposal of the same. Incineration of MSW is not suitable in India due to low calorific value of MSW. Composting of MSW is adopted in several cities and biomethanation of MSW is still in experimental stage. Hence, MSW is usually dumped in an uncontrolled manner in low lying areas in the outskirts of a city. The waste at landfill undergoes anaerobic decomposition and generates landfill gas (LFG) which is emitted into the atmosphere causing global warming due to greenhouse gas (GHG) effect. In addition, the potential hazards like fires and explosions, odour and toxicity and damage to vegetation also occur. LFG comprising of methane and carbon dioxide as major constituents, is one of the major anthropogenic sources of GHG emissions. Methane gas is twenty-one times more potent than the carbon dioxide contributing towards GHG emissions.

The atmospheric burden of methane has more than doubled during the past two centuries, and the concentration of methane in the troposphere is increasing at the rate of approximately 1% per year. Recent data indicates that global background methane levels continue to increase, but at a slower rate. The consequences of increasing atmospheric methane concentrations are several fold. Increase in tropospheric methane may lead to increase in tropospheric ozone and decrease hydroxyl radical concentrations, potentially suppressing the oxidizing potential of atmosphere and ultimately increasing the residence time of many volatile organic compounds (VOC) (Moshar *et al.*, 1999).

In India, the LFG is neither collected nor utilized till date. Hence, there is a need to determine total methane emissions from landfills in our country so as to plan and implement technologies for utilization of methane gas for useful purposes.

Intergovernmental Panel on Climate Change (IPCC) has initiated efforts globally covering inventorisation, monitoring and control of GHG emissions. NEERI, being a lead institute in waste sector under India's initial National Communication (NATCOM), carried out methane emission inventory from MSW landfills in India (NEERI, 2002).

The methodology suggested by IPCC (1996) was adopted while using default methodology (Tier I) and modified triangular approach was devised for emission inventory. The annual methane emission factors were estimated with 1994 as the base year for each State and 1999 at National level using both these methods and were estimated to be 0.0313 Gg methane/Gg of MSW and 0.0248 Gg methane emission/Gg of MSW (Year 1994) while for the year 1999, the emission factors were found to be 0.03104 and 0.0247 respectively (Kumar *et al.*, 2004).

There are a large number of uncertainties associated with the estimation of GHG emissions from landfills. These are quantity, composition of waste, extent of degradation of organic component, topography of the landfill, quantity of MSW burnt for recovery of recyclables, degree of compaction, depth of fill and age of the landfill. Hence, field studies for more number of cities (class I city and class II town) need to be carried out to arrive at more realistic estimation of methane emission factor.

NEERI carried out flux box studies to measure methane emissions at Okhla landfill site, Delhi where past records relevant to MSW were maintained. The values obtained at Okhla were compared with the values estimated using default and triangular approach. Large variation in the value of emission factor was observed. However, in the studies carried out at Bhandewadi Landfill Site near Nagpur and also at Amaravati (Maharashtra), the values obtained were 6.16 Gg/yr, 1.81 Gg/yr and estimated values using default and triangular approach were 7.65 Gg/yr, 6.97 Gg/yr for Nagpur and 2.28 Gg/yr and 1.81 Gg/yr for Amaravati respectively. These values as computed are comparable (NEERI, 2005). However, to get more realistic values, additional studies at landfill sites in other cities having past records, need to be carried out.

USEPA has reported Global human made CH_4 budget (2000) as 13% for SW amongst other sources such as enteric fermentation, manure, natural gas, oil, waste water, biomass burning, etc. Landfill methane emissions for the year 2000 as reported in global landfill methane indicated India's contribution of 1% towards methane. In 1999, LF methane emission was 20 to 70 Tg/yr out of the total emissions of 360 Tg/yr world over (Fadel and Massoud, 1999).

GHG Reduction

A number of countries have developed plans to reduce GHG emissions, and it is apparent that methane emissions can be stabilized and even reduced by adopting technological options. Calculations indicate decrease in anthropogenic methane emissions of the order of 10-20% which can stabilize atmospheric concentrations at 1990 levels (Hogan *et al.*, 1991).

Kyoto Protocol is for member countries whose objective is to reduce GHG. Clean Development Mechanism (CDM) allows governments or private entities in rich countries to set up emission reduction projects in developing countries. In addition, a key feature of protocol is the agreement on emissions trading. Under this agreement, 'holder' should be allowed 'credit' for the emission of a specified amount of GHG.

MSW (Management and Handling) Rules, 2000 recommended conversion of open dump sites to sanitary landfills for reducing methane emissions. However, this is constrained by limited budget, inability to assess methane potential, technological issues, lack of financial institutions understanding of methane projects, limited municipal capacity to develop bankable projects, limited expertise in the country to build and operate sanitary landfills and methane recovery projects. LFG Energy projects based on collection and utilization of LFG reduce methane emissions and local air pollution, create job opportunities, generate revenues and result in cost savings.

Issues for Consideration

- Adequate data is required to be generated on degradation rates of different wastes under actual field conditions. Landfills with different climatological and geological conditions need to be investigated.
- There is a need to assess and establish whether existing landfills can be improved at a reasonable cost so as to minimize GHG emissions.
- Also there is a need to determine methane-oxidising capacity from different landfills so as to decide as to whether these capacities would be enough or not to take care of most of the methane that is produced inside the landfills. In addition, more information is needed about the establishment of methane-oxidising organisms in different types of cover materials at the landfills (Borsesson *et al.*, 2000).

REFERENCES

NEERI, 2005. Status of Methane Emission from Municipal Solid Waste Disposal Sites in Nagpur and Adjoining City(s), Final Report.

Kumar, S., Gaikwad, S.A., Shekdar, A.V., Kshirsagar, P.S. and Singh, R.N., 2004. Estimation Method for National Methane Emission from Solid Waste Landfills, *Atmospheric Environment*, 38, 3481-87.

Kumar, S., Gaikwad, S.A., Devotta, S. and Singh, R.N., 2004. Qualitative Assessment of Methane Emission Inventory from Municipal Solid Waste Disposal Sites: A Case Study, *Atmospheric Environment*, 38, 4921-29.

NEERI, 2002. GHG Emission Inventory in Waste Sector, Final Report.

Borsesson, G., Galle, B., Samuelsson, J. and Svensson, B.H., 2000. Methane Emissions from Landfills; Options for Measurement and Control, Paper presented at Waste 2000 Conference, Stratford-upon-Avon, 2-4 October, 2000, 31-40.

Fadel and Massoud, 1999. Comparative Assessment of Methodologies for Methane Emissions Estimation from MSW Landfills, Proceedings Sardinia, Seventh International Waste Management and Landfill Symposium, 4-8 Oct. 1999, Italy.

Moshar, B.W., Czepiel, P.M., Harris, R.C., Shorter, J.H., Kolb, C.E., Mcmanus, J.B., Allwine, E. and Lamb, B.K., 1999. Methane Emissions at Nine Landfill Sites in the Northeastern United States, *Environmental Science and Technology*, 33, 2088-94.

Houghton, J.T. *et al.*, 1996. IPCC Guidelines in GHG Inventory Estimation, 1996, Reporting Instructions, IPCC/OECD/IEA, Paris, France, Vol. 1.

Hogan, K.B., Hoffman, J.S. and Thompson, A.M., 1991. *Nature*, 18, 354.

HAZARDOUS WASTE

27

Hazardous Waste Management and Handling Rules

T. Chakrabarti

Preamble

Growth, equity and sustainability are the three essential parameters of a development process though balancing the stakes while designing and implementing such a process is a difficult task. Adequate legislation and its implementation can only lead to an ideal development process. Ever since the Stockholm Conference on Human Environment in June, 1972, there has been a great proliferation of legislation and regulations in India to protect the environment in response to the challenge of ensuring sustainable development. This economic growth is due to a spectacular rise in industrialization throughout the country. During the last 30 years, the Indian industrial sector has quadrupled in size. However, this rapid industrialization and lax enforcement of anti-pollution laws and regulations have had a detrimental effect on India's fragile ecosystem. This detrimental effect is mainly due to the indiscriminate dumping of hazardous waste.

The main source of hazardous waste generation and impact on the environment is the chemical industry (TERI, 2003). There has been a significant increase in the number of tanneries and units manufacturing pesticides, drugs and pharmaceuticals, textiles, dyes, fertilizers, paints, chlor-alkali, etc. which have a major potential for generating hazardous wastes such as heavy metals, cyanides, pesticides, complex aromatic compounds (such as PCBs), and other chemicals, which are toxic, flammable, reactive, corrosive or have explosive properties. Regulations to control and manage air and water related pollution were in place as early as 1974 and 1981, when the Water Act and Air Act, respectively, were introduced. However, the concern and need to manage hazardous wastes arose only in the mid-eighties after the Bhopal Gas Tragedy on 2/3 December in 1984. The Ministry of Environment and Forests (MoEF) enacted the Environment (Protection) Act, in 1986, and an amendment to it in 2002. Subsequently, the MoEF promulgated the Hazardous Wastes (Management and Handling) Rules in 1989 and amendments in 2000 and 2003, and efforts to draw up inventories of hazardous waste generated in the country were initiated. Though the Hazardous Waste Rules were introduced in 1989, their implementation has remained very poor; the situation having worsened by a liberalized industrial policy that has resulted in an increase in the pace of industrialization. In this background, by order dated 13th October, 1997, a High Power Committee (HPC) with Prof. M.G.K. Menon as its Chairman was constituted by the Supreme Court to examine all matters in depth relating to hazardous waste in the country and to give a report and recommendations at an early date. The High Powered Committee so formed comprised of experts from different disciplines and fields. The High Power Committee submitted its Report on 20th April, 1998.

It appears from the HPC Report that about 80% of country's hazardous waste is generated in the States of Maharashtra, Gujarat, Tamil Nadu and Andhra Pradesh. The Report had highlighted the industrial operations (solid, liquid, gaseous waste) which result in generation of the hazardous wastes including industries recycling hazardous waste and others as detailed in the scope of work. The HPC has concluded that the hazardous wastes situation in India is fairly grim.

On the basis of the recommendations of High Power Committee, Supreme Court had passed an order on October 14, 2003 and constituted a Supreme Court Monitoring Committee (SCMC) to oversee the compliance of law, directions of this Court and Rules and Regulations and to ensure that the aspects to which the Ministry has agreed are implemented in letter and spirit and without any laxity or delay in the matter. The Monitoring

Committee is required to file quarterly reports in the Supreme Court.

Hazardous Waste (Management & Handling Rules), 2003

The Hazardous Wastes (Management & Handling) Rules, as amended on May 20, 2003, have come into force from the date of their publication in the Gazette of India, Extraordinary, Part II (3) (ii) No. 471 dated May 23, 2003. The HWM Rules, 1989 provide for the control of generation, collection, treatment, transport, import, storage and disposal of wastes listed in the schedule annexed to these rules. The rules are implemented through the State Pollution Control Board (SPCBs) and Pollution Control Committees (PCCs) in the states and union territories. The highlights of the Rules are as under:

"Hazardous Waste", as defined in the HWM Rules, means any waste which by reason of any of its physical, chemical, reactive, toxic, flammable, explosive or corrosive characteristics cause danger to health or environment, whether alone or when in contact with other wastes or substances and shall include:

- Wastes listed in column (3) of Schedule 1.
- Wastes having constituents listed in Schedule 2 if their concentration is equal to or more than the limit indicated in the said Schedule.
- Wastes listed in Lists A and B of Schedule 3 (Part-A) applicable only in case(s) of import and export of hazardous wastes in accordance with Rules 12, 13, 14 if they possess any of the hazardous characteristics listed in Part-B of Schedule 3.

The Rules do not apply to the following:

- Wastewater and exhaust gases as covered under the provisions of Water (Prevention and Control of Pollution) Act, 1974 and the Air (Prevention and Control of Pollution) Act, 1981.
- Wastes arising out of the operation from ships beyond 5 kms as covered under the provisions of the Merchant Shipping Act, 1958 and the rules thereunder.

- Radio-active wastes covered under the provisions of the Atomic Energy Act, 1962 and rules thereunder.
- Bio-medical wastes covered under the Bio-medical Wastes (Management & Handling) Rules, 1998.
- Wastes covered under the Municipal Solid Wastes (Management & Handling) Rules, 2000.
- The lead-acid batteries covered under the Batteries (Management & Handling) Rules, 2001.

Responsibility of the Occupier

As per Rule 4, it shall be the responsibility of the occupier and the operator of a facility to take all steps to ensure that the wastes listed in schedules—1, 2 and 3 of the HW Rules are properly handled and disposed off without any adverse effects to the environment. The occupier will also take adequate steps while handling the hazardous wastes to:

- contain contaminants and prevent accidents and limit their consequences on human beings and the environment; and
- provide persons working on the site with information, training and equipment necessary to ensure their safety.

If covered under Schedule 1 or 2, the industry is required to:

- apply in the prescribed format to obtain an 'Authorisation' for proper treatment and disposal of hazardous waste(s) and comply with the conditions specified in the authorization.
- take steps, wherever feasible, for reduction, recovery and recycling of wastes.
- ensure proper collection, reception, treatment, storage and disposal of hazardous wastes.
- apply for renewal of authorization before expiry of the validity period.
- maintain records of hazardous wastes handling (Form 3).
- submit "Annual Returns" to the SPCB (Form 4).
- report to the SPCB for any accident.

- conduct labeling/packaging/transportation of hazardous waste as per Motor Vehicles Act, 1988 and Rules, 1989.

Renewal of authorization is subject to the following:

- on submission of annual return in the prescribed format.
- on steps taken, whenever feasible, for reduction in the wastes generated or recycled/reused.
- on fulfillment of conditions regarding environmentally sound management of wastes.
- on remittance of processing or analysis fee.

The industries not complying with the Directions of the State Pollution Control Boards (SPCBs) can be fined as per Rule 16(3) of the HW Rules.

Packaging, Labeling and Transport of Hazardous Wastes

The compliances on the part of the operator for packaging and labeling of hazardous waste and for following the manifest system are shown below:

- The occupier or operator of a facility shall ensure that the hazardous wastes are packaged, based on the composition in a manner suitable for handling, storage and transport and the labeling and packaging shall be easily visible and be able to withstand physical conditions and climatic factors.
- Packaging, labeling and transport of hazardous wastes shall be in accordance with the provisions of the rules made by the Central Government under the Motor Vehicles Act, 1988 and other guidelines issued from time to time.
- All hazardous waste containers shall be provided with a general label as given in Form 8 of HW Rules.
- The occupier shall prepare six copies of the manifest in Form 9 of HWM Rules comprising of colour code indicated below (all six copies to be signed by the transporter).

Copy number with colour code	Purpose
Copy 1 (white)	to be forwarded by the occupier to the State Pollution Control Board or Committee.
Copy 2 (yellow)	to be retained by the occupier after taking signature on it from the transporter and rest of the four copies to be carried by the transporter.
Copy 3 (pink)	to be retained by the operator of the facility after signature.
Copy 4 (orange)	to be returned to the transporter by the operator of facility after accepting waste.
Copy 5 (green)	to be returned by the operator of the facility to State Pollution Control Board/Committee after treatment and disposal of wastes.
Copy 6 (blue)	to be returned by the operator of the facility to the occupier after treatment and disposal of wastes.

- The occupier shall forward copy number 1 (white) to the State Pollution Control Board or Committee and in case the hazardous waste is likely to be transported through any transit state, the occupier shall prepare an additional copy each for such state and forward the same to the concerned State Pollution Control Board or Committee before he hands over the hazardous waste to the transporter. Transporter shall not accept hazardous wastes from an occupier for transport unless it is accompanied by copy numbers 2 to 5 of the manifest. The transporter shall return copy number 2 (yellow) of the manifest signed with date to the occupier as a token of receipt of the other four copies of the manifest and retain the remaining four copies to be carried and handed over to respective agencies as specified in sub-rule (4).
- In case of transport of hazardous wastes to a facility for treatment, storage and disposal (TSDF) existing in a state other than the state where hazardous wastes are generated, the occupier shall obtain 'No Objection Certificate' from the State Pollution Control Board or Committee of the concerned State or Union Territory Administration where the facility is existing. A state producing hazardous waste less than 20000 mTa may send its waste to a TSDF located in another state.

- The occupier shall provide the transporter with relevant information in Form 10 of HW Rules, regarding the hazardous nature of the wastes and measures to be taken in case of an emergency.

Procedure for Registration of Re-processors of Non-ferrous Scrap and Used Oils

As per HWM Rules, 2003, waste oil (Schedule 6) used oil (Schedule 5) and other wastes listed in Schedule 4 are required to be sold or auctioned only to those recyclers possessing environmentally sound technology (EST) and registered with the Central Pollution Control Board (CPCB). Rule 21 provides for the technology and standards for re-refining/recycling of used oil/ waste oil in an environmentally sound manner. EST's stipulated in the HW Rules are as under:

- Vacuum distillation with clay treatment.
- Thin-film evaporation process.
- Vacuum distillation hydrotreating.
- Any technology approved by the MoEF.

The approved technologies and standards have distinctive advantages such as:

- The re-refining/recycling is done without adding any chemical, acids or other substances, which can increase the quantity or toxicity of waste generated in the whole process.
- The quality of products are of accepted standards for the purposes for which they are intended.

Guidelines

The Guidelines available with CPCB/SPCBs/PCCs on hazardous waste management are as under:

- Guidelines for Conducting Environmental Impact Assessment: Site Selection for Common Hazardous Waste Management Facility.
- Guidelines for setting up of Operating Facility for Hazardous Waste Management.
- Guidelines on Hazardous Waste Incinerators.

- Guidelines on the Criteria for Secure Landfill.
- Guidelines on the Upkeep of the Disposal Facility.
- Guidelines for the Transportation of Hazardous Waste.
- Pre-requisites for Hazardous Waste Authorization.

Conventions and Treaties

As a signatory to the three conventions on hazardous chemicals and waste, the Basel Convention on the Control of Transboundary Movement of Hazardous Waste and their Disposal, the Rotterdam Convention on the Prior Informed Consent Procedure for Certain Hazardous Chemicals and Pesticides in International Trade and the Stockholm Convention of Persistent Organic Pollutants, India has international obligations to effectively manage the hazardous chemicals and waste within and leaving its geographic boundaries. The Hazardous Substances Management Division (HSMD) of the Ministry of Environment and Forests (MoEF) has the responsibility of promoting safe management and use of hazardous substances, including hazardous waste.

Market

The total Indian market for environmental technologies and services is forecasted to grow from $2.6 billions in 2002 to $9 billions in 2010. The market share for hazardous waste management technologies and services was worth approximately $52 millions in 2003 and is forecasted to grow by 7% annually until 2010 (Hazardous Waste Management and Treatment Profile—India March, 2005, International Trade Canada—Trade Commissioner Service and http://www.globe-net.ca/market_reports/index.cfm?ID_ Report=435).

Conclusions

The industry driven economy of India has resulted in hazardous waste problems, which are difficult to manage in an environmentally friendly manner. The non-enforcement of 'Polluter Pays' principle, continuation of import of hazardous wastes despite the ban, absence of proper infrastructure viz. centralized disposal facilities (TSDF) and lack of technical and financial resources have led to the unscientific disposal of hazardous wastes posing serious threat to human, animal and plant life. A High Power Committee (HPC) on hazardous waste management, constituted by the Hon'ble Supreme Court of India in 1997, made similar observation and concluded that the hazardous wastes situation in India is fairly grim. Thus, there is an urgent need for formulating proper hazardous waste management strategies, implementation of hazardous wastes management regulations and establishment of proper hazardous waste storage treatment and disposal facilities (TSDFs) for controlling the unscientific disposal of hazardous wastes. This is now being done under the supervision of the Supreme Court Monitoring Committee in accordance with the order of the Supreme Court which was issued on October 14, 2003.

Biomedical Waste—Rules, Inventorization and Disposal

28

V.P. Deshpande and T. Chakrabarti

Biomedical Waste (Management and Handling) Rules, 1998

With a view to ensure proper management of biomedical wastes from health care units, the Ministry of Environment & Forests, Government of India through its Notification dated 20th July, 1998 has notified the rules called the Bio-medical Wastes (Management and Handling) Rules, 1998 (BMW Rules). These rules apply to all persons who generate, collect, receive, store, transport, treat, dispose, or handle biomedical waste in any form. The bio-medical wastes include waste from hospital, nursing home, clinic, dispensary, veterinary institution, animal house, pathological laboratory, and blood bank. BMW rules necessitate handling of wastes without any adverse effect to human health and the environment.

Under these rules, provisions have been made for making every occupier or operator to submit annual report to the prescribed authority in Form II by 31st January every year with a view to disclose information about the categories and quantities of biomedical wastes handled during the preceding year. The prescribed authority has to send this information in a complied form to Central Pollution Control Board (CPCB) by 31st March every year.

The rules have been spread over 5 different schedules with special provisions made under each schedule as given below:

> **Schedule I** specifies 10 different categories of bio-medical wastes comprising human anatomical waste, animal waste, microbiology and biotechnology waste, waste sharps (needles, blades, etc.), discarded medicines and cytotoxic

drugs, soiled wastes containing items contaminated with blood and body fluids, dressings, etc., solid waste generated from disposable items such as tubings, catheters, intravenous sets, liquid waste from labs and washings, cleaning, etc., ash from incineration of bio-medical waste and chemical waste such as disinfectants.

Schedule II deals with the colour coding and type of container for disposal of biomedical wastes. Here, treatment options for different categories of wastes mentioned in schedule I have been indicated.

Schedule III specifies labels for biomedical waste containers/ bags for exercising due care while handling the waste.

Schedule IV delineates specifications for Labels for transport of biomedical waste containers/bags.

Schedule V describes the Standards for treatment and disposal of biomedical wastes. Standards have been prescribed for operational aspects and emission standards for incinerator, waste autoclaving, liquid wastes generated in a hospital, microwaving and deep burial.

Schedule VI delineates programmed schedule for implementation of waste treatment facilities like incinerator, autoclave and microwave system.

Amendment, 2000

The Biomedical Wastes (Management and Handling) Rules, 1998 were amended (Amendment No. 375) by the Ministry of Environment & Forests, Government of India in the year 2000 on 2nd June. Under this amendment, the Central Government has made special provisions for the safety of public and health care workers from the risk arising out of biomedical waste exposure. Penalties have been stipulated for defaulters which are same as specified in Environment (Protection) Act, 1986.

Under amended Schedule I of the rules, the CPCB has recommended two types of incinerators, (1) incinerators for individual hospitals, nursing homes and medical establishments and (2) common incinerator as a centralized facility to handle waste from a number of hospitals, nursing homes and pathological laboratories. The Central Government has also indicated that the incinerators should be installed at appropriate locations to avoid nuisance to public in the neighbourhood.

Inventorization

The biomedical wastes are classified into two major categories, viz. infectious and non-infectious. The infectious wastes are capable of transmitting disease to human beings on exposure. These wastes comprise cultures and spatulas of infectious agents from lab, waste from surgical operation theatres and autopsies on patients with infectious diseases. These also include human tissues, body organs, items contaminated with body fluids, dressing materials and items contaminated with blood. Non-infectious waste is generated from packaging, food preparations and patient's activities. The quantum of infectious waste is more as compared to non-infectious waste.

A study carried out in Indore city by Malviya (1999) for establishing biomedical waste classification and proportion of various types of wastes revealed the following:

Waste type	Range (%)	Average (%)
General waste	50.46–88.31	71.37
Infectious waste	10.00–36.23	18.83
Pathological waste	4.06–19.71	8.11
Chemical waste	0.22–2.77	0.91
Waste sharps	0.41–1.42	0.78

The above proportions indicate that the percentage of constituents depends on types of facilities provided in a particular healthcare unit.

During inventorization, type of healthcare unit i.e. only consultation, with wards, Intensive Care Units (ICU) and operation theatre, etc. and sanctioned bed strength need to be known. The healthcare unit is associated with specific wards, an operation theatre and outpatient department. Separate operation theatres for ophthalmology, plastic surgery, general surgery, orthopedic, ENT and gynaecology are generally provided. In addition, there are Out-Patient Departments (OPD) as also kitchen and restaurants for serving food to the admitted patients. All these facilities generate waste.

Quantity of Biomedical Waste

In order to prepare a well-planned waste management system, it is essential to know the quantity of waste generated in the establishment. Hence, the quantities of different categories of waste generated in a hospital have to be estimated by discussions, interviews and by physical methods. The quantities generated in hospitals vary from hospital to hospital and depend on the type of health care facility, economic conditions and the category of the health care establishment. Normally, the data on waste quantities is not maintained by the hospitals. In one of the studies carried out in Indore City, the waste quantities were physically weighed in different hospitals having specialised units. On an average, the quantity of biomedical waste generation has been reported as 1.6 kg/bed/day. In India, many of the hospitals are over-occupied than the designed capacity and hence the waste quantities are derived assuming 100% bed-occupancy.

Studies carried out by NEERI in 2002 at Government Medical College and Hospital (GMCH), Nagpur indicate that there were 2599 OPD patients and 1397 admitted patients generating on an average 52.02 kg/day and 545.87 kg/day of waste respectively from OPD and inpatient sections. The waste generation rates were observed to be 0.02 kg/patient/day and 0.391 kg/patient/day respectively for OPD and inpatient sections.

In India, the waste generation rate ranges between 0.5-2.0 kg/bed/day in different healthcare units. It is estimated that annually about 0.33 million tonnes of biomedical wastes are generated in the country (Patil and Shekdar, 2001).

Handling Methods

In India, normally the waste is collected in open containers without disinfection. Bandages, cotton and other items used to scrub body fluids are collected in plastic or other non-specified containers. Waste is collected in mixed form. Some hospitals in the country are adopting colour coding as per their own norms (CPHEEO, 1998). Waste sharps are discarded without disinfection and mutilation, which may result in reusing them and spreading an infection. The waste collection and transportation workers in the hospital segregate the recyclable material for sale. Similarly all disposable plastic items are segregated by the wastepickers, from where the waste is deposited either inside the hospital campus, or outside at the community bin for further transportation and disposal along with municipal solid waste (Rao, 1995).

Treatment and Disposal

In large hospitals, normally disinfection has been practiced for treatment of infectious waste. Further, disinfected waste is disposed of along with general waste. Waste generated in OPD is also disposed of along with other general wastes. Wastes from operation theatres, wards and pathological laboratories are disposed of without any disinfection/sterilisation. Only amputated body parts, anatomical wastes, and other highly infectious wastes are incinerated wherever incinerators are available, otherwise it is burnt at some corner of the hospital campus, mostly in open pits (Patil and Shekdar, 2001).

A review of **Schedule V** of Biomedical Wastes (Management and Handling) Rules, 2000 indicate that there is no single technology to take care of all categories of biomedical waste. Hence, a hospital specific package needs to be evolved for treatment and disposal. Small and medium hospitals can have inhouse facilities such as disinfection, mutilation, shredding and autoclaving with off-site incineration at a centralized place followed by disposal in sanitary and secure landfills. Various options available for the treatment and disposal of biomedical waste are incineration, autoclave treatment, hydroclave treatment, chemical disinfection, sanitary landfill for disposal of incineration ash as also autoclaved waste and secure landfill for disposal of sharps and biomedical waste.

Swiss Centre for Development Cooperation in Technology and Management (SKAT), have compiled case studies from various developing countries like Ghana, India, Nepal, Palestinian Territories, Senegal and Tanzania, and presented in question and answer form. The authors have provided their specific comments, which would ultimately help in analysing the situation in right perspective. Finally, the lessons learned from earlier studies have been summarised. It has been stressed that management of health care waste is not simply a matter of data and technology. Adequate training, commitment, management and leadership and effective legislation are needed for better healthcare waste management (Code and Christen, 1999).

REFERENCES

NEERI, 2002. Assessment of Existing Waste Management in Healthcare Units at Government Medical College and Hospital, Nagpur, Report submitted to World Health Organization (WHO), South East Asia Region.

Patil, A.D. and Shekdar, A.V., 2001. Health Care Waste Management in India, *Journal of Environmental Management*, 63, 211-20.

Code and Christen, J., 1999. How Are We Managing our Healthcare Wastes? SKAT, St. Gallen, Switzerland, ISBN3-908001-86-2.

Malviya, K., 1999. Existing Solid Waste Management from Hospitals, Dissertation submitted to Devi Ahilya University, Indore for Award of Degree of M.Sc. in Environmental Science, 1-39.

CPHEEO, 1998. Hospital Medical Waste Management Practices in Mumbai Industrial Safety Chronicle, 77.

Rao, H.V.N., 1995. Disposal of Hospital Wastes in Bangalore and their Impact on Environment, Third International Conference on Appropriate Waste Management Technologies for Developing Countries, Nagpur, 839-42.

Inventorization and Characterization of Hazardous Waste

29

T. Chakrabarti

Inventorization

The main source of hazardous waste generation in India is the chemical industry (TERI, 2003). There has been a significant increase in the number of tanneries and units manufacturing pesticides, drugs and pharmaceuticals, textiles, dyes, fertilizers, paints, chlor-alkali, etc. which have a major potential for generating hazardous waste containing heavy metals, cyanides, pesticides, complex aromatic compounds (such as PCBs), and other chemicals, which are toxic, flammable, reactive, corrosive or have explosive properties.

The strategic plan for hazardous waste management comprises of following components:

- Inventorization, categorisation and characterization of hazardous waste.
- Improvement of existing treatment and disposal practices.
- Waste minimization at source, recycle and reuse.
- Policies on disallowing/allowing new units based on imported/indigenous hazardous waste.
- Expeditious setting up of Common Hazardous Waste Treatment, Storage and Disposal Facilities (TSDFs) through a notified governmental agency.
- Augmentation of manpower and infrastructural capacity of the Board for effective monitoring.

Thus, it is fairly well established that inventorization and characterization of hazardous waste is an integral part of the strategy and is a *sine qua non* for design of treatment, storage, disposal and facilities (TSDFs) for hazardous waste management.

The hazardous waste generated in the country per annum was estimated to be around 4.4 million tonnes (1998) while, as per the estimates of Organization for Economic Cooperation and Development (OECD) derived from correlating hazardous waste generation and economic activities, nearly five million tonnes of hazardous wastes are being produced in the country annually. This estimate of around 4.4 millions MTA is based on the 18 categories of wastes which appeared in the HWM Rules first published in 1989. Out of this, 38.3% is recyclable, 4.3% is incinerable and the remaining 57.4% is disposable in secured landfills. Twelve States of the country (Maharashtra, Gujarat, Tamil Nadu, Orissa, Madhya Pradesh, Assam, Uttar Pradesh, West Bengal, Kerala, Andhra Pradesh, Karnataka and Rajasthan) account for 97% of total hazardous waste generation. The top four waste generating states are Maharashtra, Gujarat, Andhra Pradesh and Tamil Nadu (80%). On the other hand, states such as Himachal Pradesh, Jammu & Kashmir, all the North Eastern States except Assam generate less than 20,000 MT per annum. Given the wide variations in quantity and nature of waste generated across states and union territories (UTs) and also considering the wide variations in climatic as well as hydro-geological conditions in different regions of the country, the approach to waste management has to be essentially state-specific.

Consequent upon amendments made in the year 2000 and subsequently in 2003, the State Pollution Control Boards (SPCBs) and Pollution Control Committees (PCCs) are in the process of re-inventorising hazardous waste generated. The current exercise has brought to light the serious shortcomings in the earlier inventorisation. As a result, the total quantum of waste generated as well as its composition in terms of landfillable, incinerable, etc. would undergo substantial changes. Nevertheless, the geographical distribution of waste generated and its distribution amongst the states is unlikely to undergo major changes.

While it is well recognised that inventorisation has to be reviewed and updated periodically to account for growing industrialisation, it is necessary to prepare a reliable inventory as this forms the basis for formulating a suitable hazardous waste management strategy and developing infrastructure (treatment/ disposal facilities) for their management. While field verification supplemented by stoichiometric assessments would be the ideal way forward, reasonably reliable estimates can be made based on product-wise waste streams generated and quantities thereof. In India, there are over 13,000 industrial units located in 340 districts, out of which nearly all units have been granted authorization for multiple disposal practices encompassing incineration, storage, land disposal and other disposal (mostly recycle and reuse) options.

Small and medium sized enterprises (SMEs), however, are the major hazardous waste generators.

The amount of hazardous waste generated in this country is quite small in comparison to that of the USA, where as much as 275 million tonnes of hazardous waste is generated annually. However, considering the fragile ecosystem that India has, even this low quantum of hazardous wastes (around 4.4 million MTA) can cause considerable damage to natural resources, if untreated before release. India's fragile ecosystem could be seen from the following:

- Air pollution in Indian cities is highest amongst the world.
- Over 70% of the country's surface water sources are polluted and, in large stretches of major rivers, water is not even fit for bathing.
- India has among the lowest per capita availability of forests in the world, which is 0.11 ha as compared to 0.50 ha in Thailand and 0.8 ha in China.

The security of Indian fragile ecosystem, therefore, warrants sustainable consumption of natural resources and protection from environmental degradation.

Significance of SMEs in Industrial Output and Hazardous Waste Generation

Nearly 50% of the total industrial output in India is contributed by the SMEs. They also account for 60 to 65% of the total industrial pollution. However, most of these industries generate hazardous wastes, which find their way uncontrolled into the environment. According to the National Productivity Council, New Delhi (India), there are more than 3 million small and medium scale industries, which are spread throughout the country in the form of clusters/industrial estates. SMEs in India cannot afford to adopt and maintain adequate hazardous waste treatment and disposal technologies. In the absence of common disposal facilities (TSDFs), the waste generators have been accorded temporary permissions to store waste in their premises except in areas serviced by common facilities that have come up in the States of Gujarat (9 Nos), Maharashtra (1 No) and Andhra Pradesh (1 No) where storage period should not exceed for more than 90 days. The lack of common facilities has been a major factor in mushrooming of illegal dump sites since most of the units in the small and medium sector do not have adequate space within their premises to arrange for storage over several years. Therefore, it is urgently required to make available TSDFs in the areas in all the states where SMEs are operating in large number.

There has been considerable delay in notifying sites for hazardous waste disposal. Of the 93 sites identified, only 30 have been notified. The State Governments should not only expedite notification of sites based on environmental impact assessment but play a catalytic role and persuade the industry associations to set up common facilities (TSDFs). Such common facilities would need to be planned based on reliable estimate of current waste generation and projections for the future. As this was not done, hazardous waste dumping was rampant in all the states which prompted public interest litigations in High Courts and Supreme Court.

Characterization

The Hazardous Wastes (Management & Handling) Rules, as amended on May 20, 2003 (HWM Rules, 2003), have come into force from the date of their publication in the Gazette of India, Extraordinary, Part II (3) (ii) No. 471 dated May 23, 2003. The HWM Rules, 1989 provide for the control of generation, collection, treatment, transport, import, storage and disposal of wastes listed in the Schedules annexed to these rules. The Rules are implemented through the SPCBs and in the states and union territories. According to the HWM Rules, 2003, "hazardous waste" means any waste which by reason of any of its physical, chemical, reactive, toxic, flammable, explosive or corrosive characteristics cause danger to health or environment, whether alone or when in contact with other wastes or substances and shall include:

- Wastes listed in column (3) of Schedule 1.
- Wastes having constituents listed in Schedule 2 if their concentration is equal to or more than the limit indicated in the said Schedule.
- Wastes listed in Lists A and B of Schedule 3 (Part-A) applicable only in case(s) of import and export of Hazardous wastes in accordance with Rules 12, 13, 14 if they possess any of the hazardous characteristics listed in Part 'B' of Schedule 3.

The Rules do not apply to the following:

- Waste water and exhaust gases as covered under the provisions of Water (Prevention and Control of Pollution) Act, 1974 and the Air (Prevention and Control of Pollution) Act, 1981.
- Wastes arising out of the operation from ships beyond 5 kms as covered under the provisions of the Merchant Shipping Act, 1958 and the rules thereunder.
- Radio-active wastes covered under the provisions of the Atomic Energy Act, 1962 and rules thereunder.
- Bio-medical wastes covered under the Bio-medical Wastes (Management & Handling) Rules, 1998.
- Wastes covered under the Municipal Solid Wastes (Management & Handling) Rules, 2000.
- The lead-acid batteries covered under the Batteries (Management & Handling) Rules, 2001.

The characteristics such as reactive, toxic, flammable, explosive or corrosive causing danger, etc. have not been defined in quantitative terms in the Rules. However, hazardous waste [Section 1004(7)] of Resource Conservation and Recovery Act

(RCRA) of USA defines hazardous waste as "a solid waste, or a combination of solid wastes, which because of its quantity, concentration, or physical, chemical, or infectious characteristics may: (a) cause, or significantly contribute to an increase in mortality or an increase in serious irreversible, or incapacitating reversible, illness; or (b) pose a substantial present or potential hazard to human health or the environment when improperly treated, stored, transported, disposed of, or otherwise managed". Under RCRA Section 3001, EPA is charged with defining which solid wastes are hazardous by identifying the characteristics of hazardous waste and listing particular hazardous wastes.

Current hazardous waste characteristics widely referred in USA are ignitability, corrosivity, reactivity, and toxicity which can be quantitatively expressed as under:

Ignitability

The ignitability characteristic is intended to "identify wastes capable of causing fires during routine transportation, storage and disposal, and wastes capable of exacerbating a fire once started". These risks include generally recognized fire hazards to waste management and transportation workers, such as burns and inhalation of smoke or fumes, and the potential generation and facilitated transport in air of toxic particulates and fumes that could harm the general public.

According to 40 CFR 261.21, a solid waste exhibits the characteristic of ignitability if a representative sample of the waste has any of the following properties:

Is a liquid, other than an aqueous solution, containing less than 24% alcohol by volume and has flash point less than 60°C (14°F), as determined by:

- A Pensky-Martens Closed Cup Tester, using the test method specified in ASTM Standard D-93-79 or D-93-80 (incorporated by reference, see 40 CFR 260.11),
- A Setaflash Closed Cup Tester, using the test method specified in ASTM standard D-3278-78 (incorporated by reference, see 40 CFR 260.11), or

- An equivalent test method approved by the Administrator under procedures set forth in 40 CFR 260.20 and 260.21;
- Is not a liquid and is capable, under standard temperature and pressure, of causing fire through friction, absorption of moisture or spontaneous chemical changes and, when ignited, burns so vigorously and persistently that it creates a hazard;
- Is an ignitable compressed gas as defined in 49 CFR 173.300 and as determined by the test methods described in that regulation or equivalent test methods approved by the Administrator under 40 CFR 260.20 and 260.21; or is an oxidizer as defined in 49 CFR 173.151.

Corrosivity

According to 40 CFR 261.22, a solid waste exhibits the characteristic of corrosivity if a representative sample of the waste has either of the following properties:

- Is aqueous and has a pH less than or equal to 2.0 or greater than or equal to 12.5, as determined by a pH meter using Method 9040 in "Test Methods for the Evaluation of Solid Waste, Physical/Chemical Methods," incorporated by reference in 40 CFR 260.11;
- Is a liquid and corrodes steel (SAE 1020) at a rate greater than 6.35 mm (0.250 inch) per year at a test temperature of 55°C (130°F) as determined by the test method specified in NACE (National Association of Corrosion Engineers) Standard TM-01-69 as standardized in "Test Methods for the Evaluation of Solid Waste, Physical/Chemical Methods," EPA Publication SW-846, as incorporated by reference in § 260.11.

Reactivity

"The reactivity characteristic is intended to identify wastes, which because of their extreme instability and tendency to react violently or to explode, pose a problem at all stages of the waste management process." This characteristic was intended to reduce physical risks to transportation and disposal workers and to avoid incidents that could result in the release of toxic constituents into the air consequent to an explosion or violent reaction.

40 CFR 261.23 states that a solid waste exhibits the characteristic of reactivity if a representative sample of the waste has any of the following properties:

- Is normally unstable and readily undergoes violent change without detonating;
- Reacts violently with water;
- Forms potentially explosive mixtures with water;
- When mixed with water, generates toxic gases, vapor, or fumes in a quantity sufficient to present a danger to human health or the environment;
- Is a cyanide or sulfide bearing waste which, when exposed to pH conditions between 2.0 and 12.5, can generate toxic gases, vapors or fumes in a quantity sufficient to present a danger to human health or the environment;
- Is capable of detonation or explosive reaction if it is subjected to a strong initiating source or if heated under confinement.

Toxicity

The USEPA has already predetermined that certain wastes are hazardous and these hazardous wastes have been incorporated into published lists. The hazardous waste lists are included on the VEHS website.

K-Listed Hazardous Wastes: K-listed hazardous wastes are source-specific wastes that are generated by specific industries such as iron and steel production facilities.

F-Listed Hazardous Wastes: F-listed hazardous wastes are non-specific source wastes that are generated by particular industrial processes that can occur in various industries. Industrial processes that generate F-listed hazardous wastes include wood preservation, electroplating and other metal finishing processes, and processes that generate waste solvents.

P- and U-Listed Hazardous Wastes: The P- and U-listed hazardous wastes are pure and commercial grade formulations of specific unused chemicals that are considered wastes. Unused chemicals may be considered wastes because they are no longer needed, they are spilled, or they are off-specification.

Acutely Toxic Hazardous Wastes

Certain listed hazardous wastes are considered to be acutely toxic to human health and the environment and are further defined as "acute hazardous wastes". Acute hazardous wastes include F-, K-, and P-listed hazardous wastes described above.

Conclusions

Inventorization and characterization are integral part of the hazardous waste management strategy and HMW Rules must provide characteristics of hazardous wastes in quantifiable terms so that a legally enforceable national definition of hazardous waste (in the line of RCRA, USA) can emerge. The adoption of cleaner technology and clean development mechanisms (CDMs) may soon make redundant a considerable portion of Schedule 1 of HWM Rules. Therefore, it is necessary to define hazardous waste on the basis of characteristics and not on the basis of a qualitative schedule which will result in appropriate inventorization of hazardous waste generation throughout the country.

Hazardous Waste—Treatment and Disposal

30

A.M. Deshkar

Preamble

Since World War II, accelerated industrial growth resulted in perhaps the highest standard of living in the world. However, this way of life has created environmental concerns pertaining to management of waste produced by industries and consumers.

Disasters occur due to both the natural and man-made activities (Khare, 2001). A hazard is a rare or extreme event that adversely impacts human life, property or activity. No human settlements are free from the risk of natural hazards; therefore it is vital that researchers and decision-makers have access to all available hazard information (NGDC, 2003). E&DM (2003) classified the hazards and disasters into four categories viz., natural events, technological events, man-made events and region-wise events. The adverse impacts caused due to the indiscriminate disposal of hazardous wastes (HWs) come under the category of environmental disasters. For example, in 1982, 2242 residents were evacuated after dioxin was found in soil in Missouri, U.S.A. In 1996-97, 265354 tonnes of soil and other dioxin-contaminated material from Times Beach (Missouri, U.S.A.) and 26 other sites in eastern Missouri had been incinerated (E&DM, 2003). Release of methyl isocyanate (MIC) gas in Bhopal (1984) caused a severe disaster in India (Banerjee, 2001; E&DM, 2003).

There is a growing concern all over the world for the safe disposal of HWs generated from anthropogenic sources. The hazardous waste needs to be disposed off in a secured manner in view of its strong toxic properties. Severe pollution of land, surface and groundwater may occur (Ramakrishna and Babu, 1999) if the options available (Wentz, 1995; Parsa *et al.*, 1996) for hazardous waste management are not efficiently exercised by waste generators. As per the ideal industrial siting criteria, the industries in India should have enough land available within its premises for the treatment and disposal and or reuse/recycling of the wastes generated from it (Murali Krishna, 1995). However, very few industries in India own proper treatment and disposal facilities (Jeevan Rao, 1999). Mostly the large-scale industries and a few medium scale industries (Ramakrishna and Babu, 1998), and none of the small-scale industries own hazardous waste treatment and disposal facilities. 14% of Indian industries are not having adequate facilities to comply with the regulatory standards for treatment and disposal of wastes (TEDDY, 1998).

Lack of financial, administrative and infrastructure resources are cited as some of the reasons behind these limitations. It is interesting to note that, till 1997, there was no proper secure landfill facility available in the country to dispose of hazardous waste (HPC, 2001). Hazardous wastes are required to be disposed off at treatment, storage and disposal facility (TSDF), a centralized location, catering to the hazardous waste generators from the waste generators in the near vicinity. Literature is available on the site selection and other related criteria of TSDF (Ramkrishna and Babu, 1999a; Babu and Ramkrishna, 2003). The site selection criteria for a TSDF depend upon receptors and pathways of likely waste movement, waste characteristics and waste management practices (Guidelines, 1991). The planning for hazardous waste management comprises several aspects ranging from identification and quantification of hazardous waste for development and monitoring of TSDF. In India, unauthorized dumping of hazardous waste, is however, continuing and in most of the places the hazardous waste is being utilized to fill the low-lying areas (HPC, 2001).

In a classical inventorization process, hazardous wastes are quantified based on their individual characteristics. The quantity of hazardous wastes is expressed in terms of each category for disposal (e.g. Recyclable, Incinerable or Disposable in secured landfill, etc.). The wastes that are recyclable are used/waste oil, lead wastes and zinc wastes (HPC, 2001). After quantifying the hazardous waste and assessing the probable area requirements for its treatment, storage and disposal, the sites are identified using site identification criteria as per CPCB guidelines. For this purpose, toposheets and/or remote sensing images of the study region may be used. The quantity of disposable hazardous waste (inorganic in nature to be disposed off in landfill) is high compared to the other two categories. The quantity of hazardous waste generation reported in India is 4415954 TPA from 373 districts out of 524 districts. Assuming the average density of waste to be around 1.2 tons/m^3 and the depth of the landfill as 4m, the landfill area required would be around 1.08 km^2 every year.

The demand for technically sound implementable and cost effective approaches for treating hazardous waste is of immediate and growing concern. The wastes come from many services including manufacturing processes, wastewater stream spillages and discarded products. In addition to environmental problems, these wastes represent losses of valuable materials and energy from the production processes and often require a significant investment in pollution control. Pollution prevention should be a key component of any cost effective, comprehensive waste management programme.

Hazardous Waste Prevention and Reduction Techniques

Waste prevention and reduction techniques can be divided into four major categories as shown below:

- Inventory management
- Production process modification
- Volume reduction
- Recovery

Inventory Management

The proper control of raw materials, intermediate products, and associated waste streams is an important pollution prevention technique (Hunter, 1987). The cost of disposing these materials not only includes the actual disposal cost but the cost of the lost raw materials or products. The ultimate inventory control procedures eliminate the existence of any inventory. This is done by directly moving raw materials from the receiving dock to the manufacturing area for immediate use. The final product is then shipped out without any intermediate storage.

Production Process Modification

Improving the efficiency of a production process can significantly reduce waste generation. Available techniques range from eliminating losses from process equipment to new state-of-the-art production equipment. The waste prevention and reduction techniques in this category can be divided into improved operation and maintenance, material change equipment modification and process redesign.

Volume Reduction

Volume reduction includes techniques to separate toxic, hazardous, or recoverable wastes from the total waste stream. These techniques typically are used to increase recoverability, reduce the volume and thus disposal costs. The available techniques used range from simple segregation of waste at the source to complex concentration technologies as shown in Table 1.

Table 1. Examples of Waste Reduction through Volume Reduction

Industry	Technique
X-ray film	Segregate polyester film scrap from other production waste and recyclables.
Printed circuit boards	Use filter press to dewater sludge to 60% solids and send sludge to metal recovery.
Pesticide formulation	Use separate bag houses for each process line and recycle collected dust into product.
Paint formulation	Segregate and reuse tank cleaning solvents in paint formulations.
Furniture	Segregate and reuse solvents used to flush spray coating lines and pumps as coating thinner.

Various techniques are available to reduce the volume of a waste through physical treatment. Available concentration methods include gravity and vacuum filtration, evaporation, ultrafiltration, reverse osmosis, freeze vaporization, filter press, heat drying and compaction.

Recovery

Recovering waste can provide a very cost effective waste management alternative. This technique can help eliminate waste disposal costs, reduce raw material costs, and possibly provide income from sale of the waste. Recovery of waste is a widely used practice in many manufacturing processes and can be done on-site or at an off-site facility as shown in Table 2.

Table 2. Example of Waste Reduction through Recovery and Reuse

Industry	Techniques
Mirror manufacturer system	Recover spent xylene using a batch distillation.
Printed circuit boards	Recover copper and tin or lead from process wastewater using an electrolytic recovery system.
Power tools	Recover alkaline degreasing baths using an ultrafiltration system.
Hosiery pickles	Reconstitute and reuse spent dye baths. Transfer waste brine pickle solution to a textile plant as a replacement for virgin acetic acid.
Chemical manufacture	Use spent electrolyte from one division as raw material for another.

Hazardous Waste Management Concepts

In general, hazardous waste treatment is based on chemical structure. Compounds requiring treatment can be divided into the following three categories:

- Organic, which contain carbon as derived from or found in living things or are man-made and capable of yielding electrons in oxidation reactions.
- Inorganic, which don't contain carbon as an element and are incapable of yielding electrons in oxidation reactions.
- Mixed waste, which includes both organic and inorganic constituents and is an increased challenge for treatment.

Thermal Processes

Organic compounds are combustible in air, typically have lower melting and boiling points, are less soluble in water and are biodegradable. There are many specific hazardous waste chemicals generated as a result of manufacturing processes. In many instances, the most effective way to dispose of these substances is with incineration. Some of these include the following:

- Fluids from organic acid manufacturing.
- Byproducts from pesticide manufacturing.
- Chlorosilanes from pharmaceutical manufacturing.
- Gases and liquids generated from the manufacture of vinyl chloride monomer.
- Organometallic compounds used in the glass industry.

Treatment of such wastes at the point of manufacturing is an economic and effective means of disposal. Incineration standards as per 40 CFR part 264.343 require the following:

- A 99.99% destruction and removal efficiency for each principal organic hazardous constituent (POHC).
- Hydrochloric acid (HCl) emissions, no greater than the larger of either 1.8 kg/h or 1% of HCl in the stack gas.
- Particulate matter emissions, no greater than 180 mg/dscm (0.08 grains per dry standard cubic foot) corrected to 7% oxygen.

Physico-chemical Treatment

Physical treatment processes involve physical removal of the waste from its carrier. Chemical treatment transforms the hazardous waste into a different form or compound. Physical treatment processes include coagulation, sedimentation adsorption, stripping, flotation, centrifugation, ion exchange and evaporation and encapsulation or stabilization. Chemical treatment processes include pH neutralization, precipitation, solidification and oxidation. Applicability of these physical and chemical treatment processes to various hazardous waste streams is summarized in matrix form in Table 3.

NEERI (2005) has successfully disposed off 10% arsenic containing waste stored in one of the fertilizer manufacturing industries first time in the country.

M/s. Zuari Industries Ltd. (ZIL) manufacturing fertilizers had stored arsenic bearing wastes for three decades generated during change over of the process from Giammarco-Vetrocoke (G-V) technology to Benfield technology in 1975. The studies were undertaken by NEERI between July 2004 and June 2005 which

Table 3. Applicability of Physical and Chemical Processes to Various Waste Streams

	Corrosives	Cyanides	Halogenated solvents	Non-halogenated organics	Chlorinated organics	Other organics	Oily wastes	PCBS	Aqueous with metals	Aqueous with organics	Reactives	Contaminated soils	Liquids	Solids/Sludges	Gases
Separation/filtration		X	X	X	X	X			X	X			X	X	
Chemical precipitation	X								X				X		
Chemical oxidation/reduction		X								X			X		
Dehalogenation			X		X			X				X	X	X	
Ozonation		X		X		X					X		X		X
Evaporation			X	X	X	X	X						X	X	
Solidification	X	X										X	X	X	
Photolysis													X		

included inventorization of wastes, waste characterization laboratory-scale and pilot scale solidification/stabilization studies (S/S) at ZIL, selection and assessment of waste contaminant site, design of waste contaminant facility, full-scale S/S and contaminant of waste, closure of contaminant facility and a post closure monitoring of groundwater. After the completion of S/S and contaminant operations, the contaminant tank was sealed from top by installing top HDPE liner and casting a top RCC slab. A final layer of clay and vegetative soil cover was placed on the contaminant tank.

Biological Processes

Biological processes or processes using living microorganisms have many advantages. It includes aerobic processes, anaerobic processes, composting, land treatment and enzymatic systems. Successful biological treatment can completely remove contaminants by the action of the microbes breaking down the wastes into non-hazardous byproducts. It is less expensive than many other alternatives. When complete biodegradation is not possible, removal of the hazardous compounds can be accomplished by other technologies and methods, such as activated carbon, ion exchange, and air stripping.

NEERI (2002) carried out an extensive study on "Remediation of Acrylonitrile (AN) spill sites at M/s. J.R. Enterprises (JRE), Kandla, Gujarat" during July 2001 and December 2001. The earthquake in Gujarat on January 26, 2001 resulted in cracks at the bottom of storage tank spilling over 2584 KL of AN contaminating the surrounding soil. In this study an innovative *in-situ* bioremediation route was followed which was applied in the country thereby for the first time. The laboratory isolated AN degrading cultures, *Citrobacter freundie* along with farmyard manure, was spread on contaminated soil and land was ploughed. The soil was completely remediated and AN was brought down to below detectable limit within two months from the date of application.

Infrastructure for Hazardous Waste Management

In order to organize environmental management in industrial estates, common hazardous waste treatment disposal facility

(TSDF) should be installed, commissioned, operated and managed. The sites for installing common facilities should undergo a thorough and scientific assessment. The State Industrial Development Corporation (SIDC), with its present infrastructure, cannot successfully operate and maintain such common facilities. It is, therefore, necessary to establish environmental management cooperatives with representatives from State Government, State Pollution Control Board and State Industrial Development Corporation to accomplish the tasks.

Hazardous Waste Testing Facility

Each State Pollution Control Board (SPCB) should have a hazardous waste testing facility to quantify corrosivity, reactivity, ignitability and toxicity. The capacity building exercise is being undertaken by the SPCSs with assistance of Central Pollution Control Board (CPCB) and NEERI to augment knowledge and skill base of the analysis along with infrastructure development to create hazardous waste testing facilities. The SPCB analytical laboratories must now adopt Good Laboratory Practices (GLP), a prerequisite for establishing analytical credibility of a laboratory. It is also necessary to obtain ISO 9000 series accreditation for all monitoring activities to be carried out by SPCBs.

Clearly the management of hazardous waste poses a challenge and provides an opportunity for the environmental engineering and science professions. Great effort is being expended to reduce, at its source, the quantity of waste generated and its degree of hazard by adopting cleaner processes. Looking further ahead, attention should focus on product life cycle, a systematic approach for examining the environmental and health consequences of a product at each stage of its life cycle and addressing such consequences in an integrated, cross-functional decision-making process.

REFERENCES

NEERI Report, 2005. Report on Immobilization, Solidification and Containment of Arsenic Bearing Waste at Zuari Industries Ltd., Goa.

Babu, B.V. and Ramkrishna, V., 2003. Extended Studies on Mathematical Modeling of Site Sensitivity Indices in the Site Selection Criteria for Hazardous Waste Treatment Storage and Disposal Facility, *Journal of the Institution of Public Health Engineers India*, 2003 (2), 11-17.

E&DM, 2003. Emergency Disaster Management Inc. (http://www.emergency-management.net)

NGDC, 2003. Natural Hazardous Resource Directory. (http://www.ngdc.noaa.gov/seg/hazard/resource)

NEERI, 2002. Report on Remediation of Contamination of Soil and Groundwater at Kandla, Dist. Kutch, Gujarat.

Banerjee, R., 2001. Man Made Natural Disasters, Conference on Disaster Management: Lessons Learnt from Orissa Cyclone and Gujarat — Rajasthan Drought, BITS, Pilani, 14-15.

HPC, 2001. A Report of High Powered Committee, on Management of Hazardous Wastes. (http://www.cpcb.nic.in/hpcreport)

Khare, D., 2001. Warning Precursors in Disaster Management, Conference on Disaster Management Lessons.

Babu, B.V. and Ramkrishna, V., 2000. Mathematical Modeling of Site Sensitivity Indices in the Site Selection Criteria for Hazardous Waste Treatment Storage and Disposal Facility, *Journal of the Institution of Public Health Engineers India*, 2000 (1), 54-70.

Jeevan Rao, K., 1999. Environmental Issues in Developing Countries, *Environment and People*, 6 (1), 21-23.

Lakshmi, R., 1999. Remote Sensing Application in Siting of Waste Disposal Facilities with Special Reference to Hazardous Wastes, Remote Sensing Applications in Applied Sciences, Saumitra Mukherjee (Ed.), Manak Publications Pvt., Ltd., New Delhi, 141-79.

Ramkrishna, V. and Babu, B.V., 1999. Fresh Water Depletion—A Crisis: Causes and Remedies, *Environment and People*, 6 (1), 33-39. (http:/bvbabu.50megs.com/custom.html/#23)

Ramakrishna, V. and Babu, B.V., 1999a. Site Selection Criteria for Hazardous Waste Treatment Storage and Disposal Facility, National Seminar on Waste Management, Samrat Ashok Technological Institute (Engineering College), Vidisha, India, D2-8-D2-15.

Ramkrishna, V. and Babu, B.V., 1998. Positive Decision Making in Waste Management towards Achieving Sustainable Development—A Perspective, 14th National Connection of Environmental Engineers, Institution of Engineers (India), A.P. State Center, Hyderabad, India, 9-14. (http://bvbabu.50megs.com/custom.html/#16)

TEDDY, 1998. Teri Energy Data Directory and Yearbook, 1998/99, Tata Energy Research Institute, New Delhi, 285.

Parsa, J., Stuart, H.M. and Steiner, R., 1996. Stabilization/Solidification of Hazardous Wastes Using Fly Ash, *Journal of Environmental Engineering*, ASCE, 122 (10), 935-40.

Murli Krishna, K.V.S.G., 1995. Air Pollution and Control, Kaushal & Co., Kakinada (AP), India, 307.

Wentz Charles, A., 1995. *Hazardous Waste Management*, McGraw Hill International Editions, Chemical Engineering Series, McGraw Hill Inc., 2nd edition, Singapore, 80-86, 296-351.

MOEF, 1991. Guidelines for Hazardous Wastes (Management & Handling) Rules, 1991, Ministry of Environment and Forests, India.

Hunter, J., II, 1987. Minimizing Waste by Source Segregation and Inventory Control, Technical Strategies for Hazardous Waste Prevention and Control, Sem. Government Institutes, Inc., Washington DC.

Hazardous Waste—Issues and Challenges

31

M.P. Patil

The industry driven economy of India has resulted in generation of huge quantities of hazardous wastes, which are difficult to manage in an environmentally friendly manner. The non-enforcement of 'Polluter Pays' principle, continuation of import of hazardous wastes despite the ban, absence of proper infrastructure viz. centralized disposal facilities and lack of technical and financial resources have led to the unscientific disposal of hazardous wastes posing serious threat to human, animal and plant life. A High Power Committee (HPC) on hazardous waste management, constituted by the Hon'ble Supreme Court of India in 1997, made similar observation and concluded that the '*hazardous wastes situation in India is fairly grim*' (HPC, 2001). Thus, there is an urgent need for formulating proper hazardous waste management strategies, implementation of hazardous wastes management regulations and establishment of proper hazardous waste treatment and disposal facilities (HWTDF) for controlling the haphazard unscientific disposal of hazardous waste.

Hazardous wastes are generated predominantly by the following industries: petrochemicals, pharmaceuticals, pesticides, paints and dyes, petroleum, fertilizers, asbestos, caustic soda, inorganic chemicals, metallurgical and general engineering. One of the most important issues in hazardous wastes management for such industries is lack of data/information on quantities and characteristics of hazardous wastes generated. In the absence of a legally enforceable definition and criteria for designating a waste as hazardous or otherwise, in the past (HWM Rules, 1989), the industry continued to underestimate the quantities of hazardous wastes generated by them. As a result, there is no systematic inventory on hazardous waste generation in the country. Different agencies have presented different figures on hazardous wastes generated in the country. For instance, the HPC, based on the information furnished by the State Pollution Control Boards (SPCBs) and Pollution Control Committees (PCCs), provided a figure of 4.4 million metric tonnes per annum (HPC, 2001), where as TERI (TERI, 2003) reported a figure of 7.7 million metric tonnes per annum. Moreover, huge quantities of hazardous wastes are also imported from developed countries. As reported by Ministry of Environment and Forests, Govt. of India (MoEF, 2001), 73,000 metric tonnes of hazardous wastes were imported by India during 1995 to 1997 for recycling/recovery. These include zinc ash and skimming, lead residues, lead battery scrap, copper residues and other metal bearing wastes. The quantum and disposal of hazardous wastes generated during processing of these wastes is not well documented.

Such scattered inventories of hazardous waste generation and their management were not very useful in designing hazardous waste strategies for the country since hazardous waste generation is very dynamic owing to the intense growth in industrial activities taking place. The latest rules on hazardous wastes (HWM Rules, 2003) have specified two schedules (Schedule I and Schedule II) for classification of wastes as hazardous or otherwise. The industry, with the help of these rules and in consultation with the regulatory agencies, need to prepare an inventory of hazardous wastes generated by them. The Supreme Court Monitoring Committee (SCMC) on hazardous wastes, constituted by the Hon'ble Supreme Court of India, has taken an initiative in this regard and directed the pollution control boards and committees of various States and Union Territories to prepare inventories of hazardous wastes generated under their jurisdiction. As per the latest estimates (CPCB, 2005) the total quantum of hazardous wastes generated from 34 States in India is 4.47 million tonnes per annum, which is generated by about 26,566 industry units. Among the 34 States, the major generators of hazardous wastes are Maharashtra (14,07,484 MTPA), Gujarat (12,07,000 MTPA),

Andhra Pradesh (5,07,046 MTPA), West Bengal (2,36,449 MTPA), Rajasthan (1,83,737 MTPA), and Madhya Pradesh (1,82,590 MTPA), which contribute about 83.16% of the hazardous waste generation in the country. Out of these 4.47 million tonnes of hazardous wastes, about 38% is recyclable, 4% is incinerable and 58% is land disposable. The correct figures on hazardous wastes management will help in formulating proper hazardous waste management strategies, and creating proper infrastructure for their safe and scientific management.

Another important issue on hazardous waste management for the industries is improper storage, handling, transportation, treatment and disposal of hazardous waste. Till recently, apart from few dedicated facilities at large scale industries, there were no hazardous waste treatment and disposal facilities available for small and medium scale entrepreneurs in the country. In the absence of such facilities, industries continued to store hazardous wastes within their premises in improperly designed facilities or disposed off their wastes illegally in abandoned sites. As reported by CPCB (CPCB, 2005), there are 125 illegal hazardous waste dumping sites in the country which are located mostly in the industrialized States of Andhra Pradesh, Orissa, Punjab, Uttar Pradesh, Maharashtra, Gujarat and West Bengal. However, this number is likely to go up substantially as more and more illegal hazardous wastes dump sites are being discovered. The cost of assessment and clean up of hazardous waste dump site ranges from several hundred thousand rupees to tens of millions of rupees. The hydrogeologic investigations to define the extent of groundwater and soil contamination can cost from Rs. 10 lakhs to Rs. 1 crore, whereas the average cost for cleaning up and containing contaminated groundwater ranges from Rs. 225 millions to Rs. 450 millions per site.

Currently only a few States, namely Maharashtra, Gujarat, Andhra Pradesh and Madhya Pradesh have set up a centralized hazardous waste treatment and disposal facilities (CHWTDF). Till date there are 7 CHWTDF in the State of Gujarat, 2 in Maharashtra, 1 in Andhra Pradesh and 1 in Madhya Pradesh. Other states viz. West Bengal, Tamil Nadu and Karnataka are in the advance stages of commissioning of such facilities. However, due to limited capacities of these facilities and non-existence of such facilities in other States, unscientific disposal of hazardous wastes is likely to be continued for quite some time.

Since treatment and disposal of hazardous wastes and remediation of contaminated areas due to illegal and unscientific disposal of hazardous wastes is cost-prohibitive, the future policy directions in the field of hazardous waste management should emphasize on minimization of wastes. Waste minimization can be achieved through source reduction and reuse/recycle of wastes. The source reduction can be achieved by better inventory management, improving raw material quality, substitution of less hazardous raw materials, process modifications, and better operating practice. For example, Lewis and Beautement, 2002 have demonstrated the recovery of lead from the slag generated from secondary refining of lead, through process modifications. Similarly a number of studies (Wei and Huang, 2001, Kazi and Cooper, 2002, Strobos and Friend, 2004, Csicsovszki *et al.*, 2005) have reported successful reuse, recycle and recovery of hazardous wastes.

Keeping in view the difficulties in handling, treatment and disposal of hazardous wastes in the country, and the serious adverse impacts that result from improper management of such wastes, waste minimization has assumed significant importance in India. It is, therefore, necessary to have a closer look at the processes generating hazardous wastes and explore avenues for waste minimization rather that treat, dispose and remediate hazardous waste contaminated areas. Such avenues for waste minimization would obviously be industry and product specific.

REFERENCES

CPCB, 2005. Highlights 2005, Hazardous Waste Management. (http://www.cpcb.nic.in/Highlights05/ch-8.htm)

Csicsovszki, G., Kekesi, T. and Torok, T., 2005. Selective Recovery of Zn and Fe from Spent Pickling Solutions by the Combination of Anion Exchange and Membrane Electro-winning Techniques, *Hydrometallurgy* 77, 19-28.

Strobos, J.G. and Friend, J.F.C., 2004. Zinc Recovery from Bag House Dust Generated at Ferrochrome Foundries, *Hydrometallurgy* 74, 165-71.

TERI, 2003. Hazardous Waste Management in India : A Policy Discussion Forum Base Paper. (http://www.teriin.org/seminar/waste.pdf)

MOEF, 2003. Hazardous Waste (Management and Handling) Amendment Rules, 2003. (http://www.envfor.nic.in/legis/hsm/so593e.htm)

Kazi, F.K.M. and Coope, P.A., 2002. Rapid-extraction Oxidation Process to Recover and Reuse Copper Chromium and Arsenic from Industrial Wood Preservative Sludge, Waste Management, 22, 293-301.

Lewis, A.E. and Beautement, C., 2002. Prioritising Objectives for Waste Reprocessing: A Case Study in Secondary Lead Refining, *Waste Management*, 22, 677-85.

HPC, 2001. A Report of High Powered Committee on Management of Hazardous Wastes. (http://www.cpcb.nic.in/hpcreport/)

MOEF, 2001. India, State of Environment. (http://www.envfor.nic.in/soer/2001/soer.html)

MOEF, 1989. Hazardous Waste (Management and Handling) Rules, 1989. (http://www.envfor.nic.in/legis/hsm/hsm1.html)

LAND ENVIRONMENT

Land Use and Coverage—Changing Patterns

32

Rekha Thakre

Preamble

Land and biomass are the natural interdependent ecosystems, which can be classified under nonrenewable and renewable categories respectively. Every ecosystem has a threshold for absorbing deterioration and a certain capacity for self-regeneration. Today, land environment is experiencing a tremendous stress resulting in deterioration of land quality that has sometimes crossed the resilience limits in some pockets threatening the ecological balance.

Land use may be defined as the human intervention in utilization of natural land resource for satisfying his own needs. It can be classified as crop lands, forests, waste land with and without scrubs, domestic and industrial uses, etc. The main causative factor for changing land cover is the anthropogenic intervention, due to which the natural land cover is depleting at an alarming rate. For example, the productive land in the periphery of the cities and towns is being converted into residential areas. Forests are logged off for commercial purposes without planned afforestation. Population explosion of human and livestock is aggravating the excessive pressure on crop and pastureland use.

Current Status

The total land area affected by various degrees of soil degradation through different processes is estimated at 115 Mha or 81% of the agricultural land area. Soil degradation by different processes includes 74 Mha affected by water erosion, 13 Mha by wind erosion, 7 Mha by sand dune encroachment, 3 Mha by reverines, 4 Mha by ravines and gullies, 6 Mha by poor drainage and 8 Mha by excessive salts. The total sediment loss by soil erosion in India is about 6 billion tons/yr. The rate of soil erosion ranges from 17-75 tons/ha/yr. The siltation rates of reservoirs are 3-4 times the predicted or designated rate. Urban sprawl is another factor encroaching upon arable land, and topsoil removal for brick making accounts for 0.5 to 1% of the net cropland area in some states (Swaminathan, 2000).

The per capita arable land area was 0.195 ha in 1990 and 0.176 ha in 1996. It is projected to be 0.163 ha in 2000, 0.121 ha in 2025 and 0.87 ha when the population stabilizes in 2050 or later. These projections do not consider soil degradation and conversion to non-agricultural uses.

Indo-Gangetic Plains (IGP) serve as an excellent example for studying changing land use pattern in India (Abrol and Gupta, 2000). In the last four decades, land use in the region has undergone dramatic changes. During the period between 1920s and 1940s, India had experienced a long trend of declining per capita food production. National food security for the increasing population was the main pressure leading to agricultural development since 1950s. IGP now accounts for more than half of country's net irrigated area. The irrigated areas now occupy more than 10 million ha compared to less than 4 million ha in 1960 and contribute significantly to food production. Reduction in grazing land has influenced livestock population and in turn availability of manure for maintenance of soil fertility and crop productivity. The IGP have little area under forest cover except in submontane hills of Himalayas.

Single and multiple nutrient deficiencies such as of Zn, Fe, Mn, and B have become more predominant in recent years. Long-term soil fertility studies have shown reduction in the soil's organic matter content. The biological production of soils has been gradually eroding resulting in reduced efficiency of applied inputs.

Monoculture and Bio-diversity

In the last few decades, acreage of high yielding varieties has increased substantially. A matter of serious concern is the large area coverage under a single cultivar such as HD 2329 in wheat and Jaya or IR8 in rice. Inter-cropping has been replaced by clean cultivation, reducing bio-diversity and increasing risks of build-up of new insect pest complexes as a consequence of continued cultivation of RW (Rice-Wheat) systems. Thus Phalaris minor which was only a minor weed in 1970s has assumed a menacing proportion since early 1990s. Traditional agri-silvi-horticulture has gradually given way to mono-cultures and more or less clean cultivation with fewer trees in cropped area.

Environmental Quality

Increased cropping intensity, which as a consequence of irrigation expansion particularly in the low quality aquifer zones has led to waterlogging forcing farmers to grow rice in monsoon season instead of cotton, gram, etc.

Cropping intensity was enhanced through diversification and intensification of agriculture with support from groundwater development. Altered moisture regimes resulted in oxidation of pyrites and dissolution of arsenic bearing minerals contaminating the groundwater aquifers causing serious toxicity problem. Replacement of traditional systems without anticipating the impact may well endanger millions of people as evidenced in this part of the IGP.

National Concerns

Management of land resource optimally is an important issue to be addressed. State Land Use Boards were set up in 1974 followed by The National Land Board in 1983 to fulfill the same objective. In 1988, Government of India has outlined a National Land Use Policy and Action Points for controlling the land degradation (NLUCB, 1988). However, the objective is yet to be achieved for various reasons.

Indian agriculture is presently in a dilemma on whether to increase the production at the cost of environment or to conserve the ecosystem with no concern for food security. It necessitates well thought of decisions to amalgamate the increase in agriculture production and at the same time emphasize on the scope of eco-farming at the local level for ecological connection.

Most Serious Concerns

Over exploitation and inappropriate practices are the principle factors of land degradation in India. These can be further enumerated as diversion of arable land to other uses viz. industrial development, urbanization, settlements, mining, brick-kilns, road construction and other communication modes without considering the long-term perspectives resulting in land degradation processes at an alarming rate (Yadav and Kamta Prasad, 2000). Land degradation can be measured in terms of loss of potential productivity or utility caused as a result of natural or anthropogenic factors. Mechanisms of land degradation include physical, chemical and biological processes. Important among the physical processes are decline in soil structure leading to crusting, compaction, erosion, desertification, anaerobism, environmental pollution and sustainable use of natural resources. Significant chemical processes include acidification, leaching, salinization, decrease in cation retention capacity and fertility depletion. Biological processes include reduction in total carbon and decline in land biodiversity.

The land degradation due to erosion of water is the biggest soil scourge in India. The annual recurring soil loss is estimated to be 5300 Mtons of topsoil, which not only causes huge loss of plant nutrients but also is responsible for depleting the resource annually (Kanwar, 2000).

In Northeast Himalayan region, Jhuming of short cycle (3-6 yrs now) aggravate erosion and related hazards. Illegal felling of forest, grazing pressure of animal population (there is hardly 4.1% area under pasture and grazing land) also accentuate the erosion and environmental problem including extinction of many valuable floral and faunal species.

Wind erosion is a severe threat to arid and semi-arid environments covering Rajasthan, Haryana, Gujarat and Punjab and sandy tract of coastal areas and cold desert. Terrain and soil characteristics, low rainfall, high wind velocity and sparse or absence of natural vegetation cover are responsible for accelerated wind erosion.

Also, injudicious use of NPK fertilizers without organic source lead to depletion of one or many nutrients in soil. This in turn, limits the crop production both qualitatively and quantitatively. The continuous application of N fertilizers alone causes sharp reduction in soil organic carbon (SOC) levels, a key indicator of soil quality, irrespective of the cropping system. About 70% area in country is deficient in SOC with levels of organic carbon around 1% (Velayuthum *et al.*, 1999).

It is estimated that a total of 10.1 Mha of soil has the problem of salinity/alkalinity associated with it. However, the sodicity in black soil is more detrimental than those of Indo-Gangetic plain. The brackish water associated with these areas also pose health hazards due to toxic elements like F, Se and B. Domestic, municipal and industrial wastewater and their unregulated dumping and/or unscientific use is of great concern from soil quality and health hazards point of view because they contain heavy metals like Cu, Zn, Mn, Hg, Pb, Cd, Ni, etc. and pathogens (NEERI, 2003). Land degradation due to heavy metal contamination such as arsenic, chromium, copper and lead, has been reported in Mumbai, Kanpur, Dewas (M.P.) and Chennai. The discharge of tannery and distillery effluents on land has led to degradation of soil quality in different areas of Tamil Nadu.

Inadequacy in Management and Control Practices

Prevention of land degradation and cleanup of contaminated land have become a worldwide environmental priority. This involves the concept of using "improving and restoring" the productive capacity of land. The goal should be not only to minimize soil degradation but also to protect soil quality. The land use or farming system to be adopted must be land restorative rather than land-depletive, fertility mining or soil degrading. The land should not be misused or made a dumping ground for toxic solid/liquid waste, as it has a limited built-in resistance to abuse (Lal and Stewart, 1992).

Degradation of land conditions is probably most widespread and pernicious form of deterioration, because it affects a major life supporting system and its natural recuperation may take centuries. The quality of these lands can be improved by physical, chemical and biological remediation but it requires systematic diagnosis of problems and research by a group of scientists at local level.

In addition to the environmental implications of soil degradation, there are human dimensions to this problem. There is an increasing awareness that resources degradation is also linked to poverty and there is a vicious circle linking and eventually governing these two major problems of the human society. The most difficult part is to realize the cause and effect relationship between the two and perhaps it can be best explained as the chicken and egg controversy.

Future Challenges

In India, the land degradation factors are to be assessed on a micro-scale using sophisticated techniques like remote sensing coupled with land survey and mapping so that exact cause and effect relationship for land degradation can be established. This will help in framing the corrective action for restoration of degraded land and controlling the degradation process further. In addition, evaluation of assimilative capacity of land according to their designated uses at micro- and meso-scale is the need of the hour.

Disposal of urban and industrial effluents on land is mostly practiced in crop production in India as well as other developed and underdeveloped countries. Judicious reuse of waste waters and recycling of plant nutrients in agriculture combines the advantages of augmenting the irrational and manurial resources of the country for the development of agriculture and simultaneous control of the related environmental pollution problems. However, if it is disposed off without any consideration, it becomes a liability to the nation as it involves a considerable degree of treatment prior to its conventional disposal in surface waters and also involves a tremendous cost. Unscientific and unsystematic land disposal of wastewater poses adverse impact on soil texture, quality and productivity. Nevertheless, in India there are no guidelines/technology for land application of wastewater for crop production which warrant urgent attention.

Assessment of Relative Pollution Hazard

In general, heavy metal concentrations in soil increase with increasing proximity to industries and intensity of human habitation. Metal concentration in most soils remains below critical limits for agriculture, but may exceed if accumulation occurs in topsoil. Enrichment of heavy metals in soil as a result of solid and liquid waste has to be considered from assimilation capacity point of view. Due to land application of solid and liquid wastes, there is a potential risk of groundwater pollution. This aspect needs to be considered judiciously. Specific Organic and Inorganic Pollutants are to be given due attention during the assimilative/carrying capacity studies at regional/local level.

REFERENCES

NEERI, 2003. Use of Urban and Industrial Effluents for Crop Production, Sponsored by NATP-ICAR.

Abrol, I.P. and Gupta, R.K., 2000. Indo-Gangetic Plains—Issues of Changing Land Use, LUCC Newsletter 3.

Kanwar, J.S., 2000. Soil and Water Resource Management for Sustainable Agriculture, International Conference on Managing Natural Resources in the 21st Century, New Delhi, February, 4-6.

Swaminathan, M.S., 2000. International Conference on Managing Natural Resources in the 21st Century, New Delhi, February 14-15, 5-14.

Yadav, R.L. and Kamta Prasad, 2000. Modern Concepts for Ecofriendly Management of Cropping System, International Conference on Managing Natural Resources in the 21st Century, New Delhi, February, 4-6.

Velayutham, M., Mandal, D.M., Mandal, C. and Sehgal, J., 1999. *Agro-ecological Sub-regions of India for Developmental Planning*. NBSS&LUP Publication, 35, NBSS&LUP, Nagpur, India, 452.

Food and Agriculture Organization, 1992. Wastewater Treatment and Use in Agriculture, FAO, Rome.

Lal, R. and Stewart, B.A. (Ed.), 1992. *Advances in Soil Science*, Springer-Verlag, New Delhi, 17.

National Land Use and Conservation Board (NLUCB), 1988. Report of the Committee of Experts on Draft Outline of National Land Use Policy, Department of Agriculture and Cooperation, Ministry of Agriculture.

Gupta, D.K., 1986. *Pesticides in Indian Environment*, Inter Print Publisher, New Delhi.

33

Topsoil

Rekha Thakre

Preamble

Soil is an important reservoir of nutrients required for plant growth, which are mostly concentrated in the top layers. Topsoil is also referred to as a living component of the earth as many microbes inhabit in this fraction and help in bio-geo-chemical cycling of carbon, nitrogen, oxygen, sulphur and water.

Soil is a natural resource and is not renewable in a short period of time. Under most natural conditions, soil is formed at a rate of approximately 1 cm in every 100 to 400 years. Soil is very expensive to reclaim or to improve once it is physically degraded or chemically depleted (Stoops and Cheverry, 1992).

Current Status

Land degradation processes are claiming a heavy toll of soil resources every year (Prasad and Biswas, 1998). The NBSS & LUP estimates of water erosion account for 132.5 Mha loss of topsoil and 16.4 Mha of soil deformation. The average annual soil loss is estimated to be about 5300 Mt of soil. Out of this, 29% of soil is lost permanently to sea, 10% deposited in irrigation reservoirs and 69% deposited as alluvium in different places. An area of 13.5 Mha land is degraded by wind erosion, 10.8 Mha by salinization and 11.6 Mha by physical deterioration such as waterlogging. The loss of plant nutrients associated with the soil is also very large.

Brick making for housing construction is worsening situation still more than degradation of soil productivity. In this activity, in addition to soil degradation, there is an additional loss as the soil used for brick making is totally lost forever. In brick kiln industries, topsoil is mainly used as a raw material on large scale. It is known that most of the plant nutrients are enriched on the top layers, which are also lost due to this activity, and hence fertility status of soil becomes low.

Based on the investigation studies at NEERI, in the states of Assam, Maharashtra, Punjab, Rajasthan, Tamil Nadu, Uttar Pradesh and West Bengal, it is estimated that about 173.35 Mtons of topsoil is lost in this activity every year in all these states. These losses are, however, highly on lower side. This loss is never included in the assessment of topsoil loss estimates. Soil loss is maximum in UP followed by Punjab. Nevertheless, in Punjab, considering the area of the state, the brick making activity is on larger scale than any other state of India. It is also noteworthy that due to the Indo-Gangetic drainage network, the states of Punjab, Uttar Pradesh and West Bengal are endowed with rich alluvial soils up to considerable depths. Nutrient loss scenario projects that brick kiln industry accounts for 245 to 2237 kg/ha, 125 to 920 kg/ha and 2360 to 14804 kg/ha for total N, total P and total K respectively every year (NEERI Reports for UP—2003, West Bengal, Punjab, Maharashtra—2003, Rajasthan—2003, Tamil Nadu—2003 and Assam—2004).

Indo-Gangetic Plains (IGP), spread over South Asia's four countries, Bangladesh, Nepal, India and Pakistan are agriculturally one of the most important regions of the world. In India alone, IGP occupies 65 Mha of area covering the states of Punjab, Haryana, Uttar Pradesh, Bihar and West Bengal, which is about 1/5 of the total geographical area of the country.

Intensive farming practices, particularly with wheat and rice have resulted in virtually mining nutrients from the soil. Also due to heavy use of fertilizers, excess nitrates have leached into groundwater leading to significant increase in groundwater contamination. Over-application of chemicals has degraded the land significantly. Apart from overuse of chemicals, equally important issue is imbalance in the application of fertilizers and pesticides.

Intensive agriculture has also led to extension of area under irrigation. In India, area irrigated has doubled over four decades, from 19% to 38% of the net sown area. Much of this increase has come from water extracted from the fast depleting groundwater resources. Improper use and maintenance of canal irrigation has contributed significantly to the soil degradation problem. Extension of canal irrigation to arid and semi-arid areas has resulted in waterlogging and salination. According to the National Remote Sensing Agency and Forest Survey of India, 60% of the total area under cultivation is substantially degraded. Most of this damage is in the form of loss of topsoil.

Jhoom cultivation practiced by tribals in the Himalayan regions of India (Northeast and West) is another causative factor in topsoil losses. The tribals in this hilly region slash and burn the forest trees for cultivation, practice agriculture for short duration and shift to other places as the productivity decreases after 3-4 years. This results in heavy losses of topsoil which is stored for ages by the tree cover on hilly slopes.

National Concerns

Loss of soil fertility resulting in low yields (in spite of applying the costly fertilizers) is a pressing problem for the farmers who are not aware of the effects of loss of topsoil due to various factors.

Topsoil losses due to water erosion result in silting of reservoirs. The life of a dam is directly affected by the rate of siltation, which is a major threat to utility of reservoir. Many major dams in India are facing eutrophication problems associated with siltation. Bhakra-Beas Management Board is working hard to stop siltation and other problems associated with it.

In many of the most productive areas, there are problems of declining soil quality that, if unchecked, can lead to permanent loss of prime agricultural land. Low profits have kept the farmers from adequately investing in resource conservation, while subsidies have caused over use of agricultural chemicals with subsequent environmental damage.

International Status

Around 1.96 billions of ha world-wide are affected by human induced soil degradation, mainly caused by water and wind erosion (1094 million ha; 548 million ha respectively). Chemical degradation accounted for 240 million ha, apart from nutrient decline (136 million ha) and secondary salinization (77 million ha). Physical degradation occurred on 83 million ha mainly as a result of compaction, sealing and crusting (Oldeman, 2000).

Most Serious Concerns

Changes in soil and terrain properties such as loss of topsoil, development of rills and gullies, and loss of nutrients may reflect the occurrence and intensity of the process of soil degradation, but not necessarily the seriousness of its impact on food productivity. The removal of 0.5 cm topsoil layer may have a serious effect on a shallow soil with a thin topsoil but may not directly impact the productivity on a deep fertile soil.

Resource inventory for carbon stock in terms of soil organic matter (SOM) of Indian soils become imperative to identify precisely and provide guidance in prioritizing the eco-regions where soils are degraded and need immediate rehabilitation to increase the soil productivity. This also in turn addresses the carbon dioxide accounting of the ecosystem, which is the prime culprit in global warming phenomenon.

Catchment areas of the river basins pose a great risk for soil losses due to water erosion and should be treated on top priority basis. These areas are mostly responsible for siltation of the down stream multi-purpose reservoirs. Once silted, the reservoir loses its utility value.

Inadequate Management and Control Practices

Most of the catchment areas are not planned and properly treated according to the land use suitability leading to shortening of the life span of the reservoirs.

SOM is the key factor in productivity of the topsoil. SOM is known to serve as soil conditioner, nutrient source and substrate for microbial activity, preserver of the environment and major determinant for sustaining or increasing agricultural productivity (Schnitzer, 1991). The current status of SOM in India indicates that more than 50% of area of the country falls under arid (hot) and semi-arid (hot) agro-ecological regions containing <2% SOM content. The soils of these eco-regions, therefore, require carbon sequestration. They are represented by the alluvial soils of Indo-Gangetic Plains, shrink-swell soils of central, western and southern India and ferruginous soils of southern peninsular regions.

To effectively develop and implement programmes on food sufficiency, land use optimization, biodiversity conservation and land degradation control are mandatory. Policy makers require information on the extent and quality of land related potentials and constraints. There is, however, a growing concern that natural resources information is underused in decision-making by policy makers, extension services, and land use managers.

Future Challenges

Soil erosion risk and its impact on food productivity is highly important and a great challenge.

The level of SOM should be maintained at an acceptable level, above critical values corresponding to structural stability and biological activity. But, no critical level is yet established. Dynamic strategies like regular inputs of SOM and inorganic fertilizers are needed to maintain and enhance the soil productivity, along with suitable cropping pattern. Agro-forestry/agro-horticulture system offers enormous scope to restore soil health by sequestration of organic carbon.

Tillage increases SOM oxidation by disrupting soil aggregates, exposing new surfaces to microbial attack and changing the redox conditions within the profile of topsoils (0-7.5 cm). It is documented that the changes in tillage and fertilizer management practices required for no-till soils reflect the increased potential for immobilization of surface applied N and the lower levels of plant available Nitrate as compared to those under conventional tillage. Adoption of no-till management practices also influence the carbon cycle in the soil (Doran, 1980).

Catchment areas of the small and large river basins need proper attention to control topsoil losses. Planning and implementation of the afforestation on the slopes, contour bunding, and agricultural practices, which enhance the soil degradation, should be meticulously formulated on micro-scale level to stop soil degradation.

Land is a non-renewable, finite resource mostly required for the sustenance of flora and fauna existing on the planet earth. If it is lost, the very existence of the bio-system will be lost.

REFERENCES

NEERI, 2003, 2004. Restoration and Remediation of Top Soil Losses due to Brick Kilns in India (Submitted to Ministry of Environment & Forests: UP—2003,West Bengal, Punjab, Maharashtra—2003, Rajasthan—2003, Tamil Nadu—2003 and Assam—2004).

Oldeman, L.R., 2000. Impact of Soil Degradation: A Global Scenario, International Conference on Managing Natural Resources for Sustainable Agricultural Production in 21st Century, 79-86.

Prasad, R.N. and Biswas, P.P., 1998. Soil Resources of India, 50 Years of Natural Resources Management Research (Singh, G.B. and Sharma, B.R. eds.), ICAR, New Delhi, 13.

Stoops, G. and Cheverry, C., 1992. New Challenges for Soil Research in Developing Countries: A Holistic Approach, Workshop of the European Community, Life Sciences and Technologies for Developing Countries (STD 3 programme), Rennes, 1992, 22.

Schnitzer, M., 1991. Soil Organic Matter—The Next 75 Years, *Soil Science*, 151, 41-58.

Doran, J.W., 1980. Soil Microbial and Biochemical Changes Associated with Reduced Tillage, *Soil Science Society of America Journal*, 44, 765-71.

NBSS & LUP Report. UNDP Technical Report on Study of Factors of Land Evaluation System, 73.

Soil Pollution

N.P. Thacker

34

Preamble

Land contamination is usually the result of previous land usage(s) or may, in certain cases, be due to contaminants being present due to natural geological conditions. The term land contamination, therefore, covers a wide range of situations where land is contaminated in some way. Often this is associated with industrial processes or activities that have now ceased, but where waste products or remaining residues present a hazard to the general environment and may represent harm to human health. The typical land pollutants include waste oils, solvents, fuels, polychlorinated biphenyls (PCBs), contaminated wastes and pesticides. (www.deq.stateor.us/wq/dwp/DWPTTypicalContaminants.pdf/)

Pollutants

In India, the major land environment pollutants include fluoride, heavy metals like iron, arsenic, lead, cadmium, mercury, pesticides like DDT, BHC, endosulfan and PCBs. In 1999, Greenpeace and Bhopal community groups documented the presence of stockpiles of toxic pesticides as well as hazardous wastes and contaminated material scattered throughout the factory site as a result of Bhopal Union Carbide disaster. This survey found substantial to severe contamination of land and water supplies with heavy metals and chlorinated chemicals. (Earth summit, 2002). (http://archive.greenpeace.org/earthsummit/news bhopalprinciplescs.html)

Heavy Metals

Heavy metal contamination viz. Mercury, strontium, zinc, cadmium, iron, etc. in Andhra Pradesh, Gujarat and Haryana is alarming. Mercury levels in drinking water were found to exceed 150-250 times the WHO prescribed GVs (Down to Earth, 1999) (http://www.rainwaterharvesting.org/Crisis/Groundwater-pollution.htm). The concentration of heavy metals in the marine sediments in Gujarat was reported by Reddy *et al.* (2004). In Andhra Pradesh State, the arsenic concentration in the soil of the 14 sites ranged from 0.87 to 12.8 mg/kg (Sekhar *et al.*, 2003). The highest concentration was found at Kazipalli, Hyderabad in 2001, 12.0±1.8 mg/g which is much above the EPA permissible limit of 20 mg/g.

The mean concentration of Heavy metals in soil samples collected from Murshidabad district, West Bengal in 2001 were found as (mg/kg)—As (5.31), Fe (6740), Cu (18.3), Pb (10.4), Ni (18.8), Mn (34.2), Zn (41.2), Se (0.53), Mg (340), V (44.6), Cr (33.6), Cd (37) (Roy *et al.*, 2002).

Among the heavy metals (HMs) levels monitored in tube wells of Delhi, lead, manganese, copper, cadmium and iron were found marginally above the BIS specifications (Dixit *et al.*, 2003), whereas recently in 2004, the HMs monitored quarterly in tube wells at Delhi were found to be below the BIS specifications except manganese and chromium which marginally exceeded the BIS specifications (DJB, Delhi Report, 2004).

Pesticides

Recent studies (Thacker *et al.*, 1997 and 2001) have confirmed widespread contamination of surface and groundwater with pesticides, confirming the existence of a problem requiring an appropriate policy response.

Lindane, endosulfan and total DDT levels were monitored in lakes, rivers and tube wells in Nagpur in 1996 and 2004 during winter and summer seasons. The DDT levels were found to marginally exceed the WHO GV (Thacker *et al.*, 1995 and 1997).

Central Ground Water Board (CGWB), in collaboration with the Central Pollution Control Board (CPCB) in April, 1999 conducted the study for the National Capital Territory of Delhi and reported presence of pesticides in 15 out of 127 samples.

Maximum total DDT concentration of 2.6 ppm was detected in the soil from Durga Nagar in the vicinity of a DDT factory (Yadav *et al.,* 1981).

Forty-five soil samples of surface (0–15 cm) and subsurface (15–30 cm) soils from agricultural sites of Delhi, Haryana, Haridwar, Uttar Pradesh (UP) and around the hexachlorocyclohexane (HCH) manufacturing plant of IPL (Indian Pesticides Limited) and nine samples of different commercial brands of drinking water from markets in Delhi were analysed for the presence of residues of HCH isomers. Thirty-nine (out of 45) soil samples contained residues of *b*-HCH (2.5 mg/kg to 463 mg/kg of soil) and the remaining showed the presence of *g*-HCH (0.08 mg/kg to 43.00 mg/kg). Residues of *a*-HCH (0.04 mg/kg to 98.00 mg/kg of soil) and *d*-HCH (0.07 mg/kg to 458.00 mg/kg of soil) were detected less frequently. Bisleri, Paras and VIP brands showed 99, 240 and 141 fold higher levels of HCH respectively when compared with the EEC norms. The contamination of mineral water could be due to heavy contamination of groundwater, which is used for making mineral water (Om Prakash *et al.*, 2004).

The concentration of *g*-HCH in soil samples from Kolkata in 2000 ranged from 0.003 to 0.33 g/g[1] and of DDT 0.003 to 0.119 g/g (Bhattacharya *et al.*, 2003). The summary of CPCB findings regarding pesticide pollution in groundwater in various Indian states is reported (www.cpcb.org, 2004).

PCBs

PCBs are environmental contaminants on global scale. PCBs are considered to constitute severe health hazard due to organic compounds normally encountered in landfill sites. In Nagpur, the study was carried out by NEERI in 2000–2001 (Thacker *et al.*, 2001) and maximum levels of total PCBs (Arochlor 1260 and Arochlor 1254) on dry weight basis were found to be 1.2 g/g to few mg/g (Thacker *et al.*, 2002).

Inadequacy in Management

- It is common knowledge in India that industrial units does not have enough resources to treat effluents and solid wastes properly and hence dump them in open or inject them directly into the ground and water.
- Most of the industries do not have facility for incineration of hazardous wastes.
- Pesticides and many toxic chemicals that have been either banned or usage restricted in many countries are liberally used in India.

Future Challenges

- Monitoring of Pollutants.
- Development of National standards/regulations and their enforcement.
- Studies on health effects.
- Integrated control measures on mitigation and prevention.
- Implementation of cost-benefit based approach.

REFERENCES

Anjaneyulu, Y. and Khan, Z., 2005. Influence of Soil Components on Adsorption Desorption of Hazardous Organics Development and Low Cost Technology for Reclamation of Hazardous Water Dump Sites, *Journal of Hazardous*, 118 (1-3), 161-69.

Gupta, A.K., 2004. Pesticide Exposure—Indian Scene, *Toxicology*, 198 (1–3), 85-90.

Om Prakash, Mrutyunjay Suar, Vishakha Raina, Charu Dogra, Rinku Pal and Rup Lal, 2004. Residues of Hexachlorocyclohexane Isomers in Soil and Water Samples from Delhi and Adjoining Areas, Current Science, 87(1).

Reddy, M., Basha, S., Kumar, V.G.S., Joshi, H.V. and Ramachandrauih, G., 2004. Distribution Enrichment and Accumulation of Heavy Metals in Coastal Sediments of Along-Sosija Ship Cropping Yard, India, *V-Marine Pollution Bulletin*, 48 (11-12), 1055-59.

Dixit, R.C., Verma, S.R., Nitnaware, V. and Thacker, N.P., 2003. Heavy Metal Contamination in Surface and Groundwater of an Urban City, *Indian Journal of Environment & Health*, 45 (2), 107-12.

Sekhar, K.C., Chary, N.S., Kamala, C.T., Venkateswara, J., Balaram, V. and Anjaneyulu, S., 2003. Risk Assessment and Pathway Study of Arsenic in Industrially Contaminated Sites of Hyderabad; Case Study, *Environment Journal*, 29 (5), 601-11.

Roy Choudhary, T., Uchino, T., Tokunaga, H. and Audo, W., 2002. Arsenic and Other Heavy Metals in Soils for an Arsenic Affected Area; *Chemosphere*, 49 (6), 605-18.

Thacker, N.P., Titus, A., Pande, S.P., 2002. Polychlorinated Biphenyls in Compost Amended Soil of a Landfill Site, *Indian Journal of Environmental Health*, 44 (1), 19-23.

Titus, A., Thacker, N.P. and Pande, S.P., 2002. Polychlorinated Biphenyls in Landfill Site of an Urban City, Indian Association for Environmental Management, 29, 66-70.

Thacker, N.P., Vaidya, M.V. and Sipani, M., 1995. Sustainability of Indian Water Systems for Organohalides, 3rd Cong. Toxicol. Dev. Count., Cairo, Egypt, Nov. 19-23, 1995, Proc: III, 213-22.

Yadav, D.V., Mittal, P.K., Agarwal, H.C. and Pillai, M.K., 1981. Organochlorine Insecticide Residues in Soil and Earthworms in the Delhi Area, India, August-October, 1974. *Pesticides Monitoring Journal*, 15(2), 80-85.

Thacker, N.P., Kaur, P. and Rudra, A., 1997. Halogenated Compounds in and Around Nagpur City Water Resources, India, *International Journal of Environmental Education and Information*, 16(4), 367-76.

STATEOR. (http://www.def.stateor.us/wf/dwp/dwpttypicalContaminants.pdf)

ENTRIX. Glossary by ENTRIX, an Environmental Consultancy. (http://www.entrix.com/resources/glossary.aspx)

Greenpeace, 2002. Greenpeace Earth Summit, Dow Bhopal (India). (http://archive.greenpeace.org/earthsummit/news_bhopalprinciples_cs.html)

CSE, 1999. An Article from Down To Earth. (http://www.rainwaterharvesting.org/Crisis/Groundwater-pollution.htm)

CPCB. (http://www.cpcb.org)

Phytoremediation

A.A. Juwarkar

35

Preamble

Phytoremediation is defined as the use of living green plants for treatment or remediation or clean up of contaminated sites or ecosystems, contaminated soil, sludge, sediments, and groundwater through contaminant removal, degradation, or containment. The technology is low cost, economically sustainable, natural and self-sustainable and is an aesthetically pleasing, solar-energy driven, passive technique. It can be used to clean up sites with low to moderate levels of contamination. It can also be used along with or, in some cases, in place of mechanical cleanup methods. Phytoremediation includes a variety of remediation techniques for the environment, which include: rhizodegradation (uptake of contaminants in plant roots), phytotransformation (plant uptake of organic contaminants and degradation), plant-assisted bioremediation (microbial degradation in the rhizosphere region), phytoextraction (direct uptake of contaminants in plant tissues with subsequent removal of the plants), phytostabilization (roots exudate cause metals to precipitate and biomass becomes less bioavailable), phytovolatilization (plants evaporate certain metal ions and volatile organics), and removal of aerial contamination (uptake of various volatile organics by leaves).

This technology has been implemented commercially as well as at pilot scale, at many sites in the U.S. as compared to other countries (Schnoor, 2002). It was estimated that the total U.S. phytoremediation market in 1998 to be $ 16.5-29.5 millions, and that market would grow to $ 214-370 millions by 2005 (Glass, 1998). In India, its implementation is still in infancy.

Current Status

NEERI, Nagpur has made efforts to rejuvenate mine spoil dumps, mine tailings, fly ash dumps, metal mine spoil dumps (Fe, Cu and Zn) using microbially assisted phytoremediation (Juwarkar *et al.*, 2004; Juwarkar *et al.*, 2003; NEERI Project Report, 2000; Juwarkar *et al.*, 1998; Juwarkar *et al.*, 1992; Juwarkar *et al.*, 1991). The phytoremediation technology has been demonstrated on commercial scale at different mining sites and has been recognized and awarded nationally and internationally. In other findings, an ecotype of the brake fern *pteris vittata* has been identified as a hyper accumulator of arsenic (10908—4700 mg As/kg dry biomass).

Aquatic vascular plants like *Hydrilla verticillata, Spirodela polyrrhiza, Bacopa monnierii, Phragmites karka* and *Scirpus lacustris* have been used to treat chromium contaminated effluent and sludge from leather tanning industries (Rai and Pal, 1999). *Brassica juncia L* is being studied for its potential to uptake metal (MoEF Annual Report, 2001-02). Phytoremediation of particulate matter from ambient environment through dust capturing plant species has been initiated at Pollution Control Research Institute (PCRI), Hardwar. Phytoremediation of fly ash dumps has been demonstrated at Badarpur near Delhi and Korba in Chhattisgarh (TERI Report, 2000). Sodic soils have been remediated by fast growing and multipurpose trees plants such as *Dalbergia sisso, Acacia nilotica* and *Azadirachta indica* to increase the productivity of the soil at Sultanpur and Allahabad districts of Uttar Pradesh (FRI Annual Report, 1977). Phytoremediation of ravine lands, saline soil and mined out areas have been carried out by plants viz. *Pongamia pinnata*, *Gravillea pteridifoli* and *Grevillea robusta* (Irshad Khan, 1987). Restoration of degraded ecosystems has been initiated at Asola and Bhatti wildlife sanctuaries (Babu, 2002). The plant *Azolla spp.* is being studied for its accumulation of heavy metals (Arora and Singh, 2004). Phytoremediation of coalmine by *Acacia nilotica, Acacia mellifera* and bauxite mined out areas by *Gravillea pteridifolia* has been demonstrated at Dhanpuri and Amarkantak in Madhya Pradesh (Ram Prasad, 1999).

International Concerns

During the last five decades, the worldwide release of heavy metals reached 22,000 t for cadmium, 9,39,000 t for copper, 7,83,000 t for lead and 1,35,000 t for zinc. International worry on phytoremediation of metal contaminated sites centers around the fear of incorporation of toxic chemicals into the food chain through grazing, etc., if these sites are not maintained properly. The other major concerns about these efforts pertain to the disposal of plants and trees used for phytoremediation of hazardous sites. Most of the plants that grow on metal contaminated soil can accumulate and store heavy metals in their roots and shoots to some extent. Metal hyper accumulator plants can accumulate and concentrate metals in their shoots to levels of at least 0.1% metal/dry biomass. Such hyper accumulators can be used to reduce inorganic contaminants to levels that are environmentally acceptable and meet regulatory requirements.

Inadequacy in Management and Control

Maintenance of the phytoremediation sites should include routine inspection of groundwater, meteorological and movement of contaminants and plants themselves. The assessment of phytoremediation and its management includes factors such as site conditions, field pilot test, sampling, monitoring, cost and performance. To manage the phytoremediation sites, the understanding of site conditions such as site layout, hydro geological settings, groundwater capture and water balance modeling, meteorological monitoring and agronomic evaluation is essential.

Regular monitoring and sampling of the phytoremediation site should be conducted to assess the movement of the contaminant. Monitoring wells and piezometers should be installed to monitor the contaminants and to evaluate the phytoremediation technology.

Emerging Issues

Limitations of phytoremediation include the difficulty of treating wastes greater than three meters deep, possible uptake of contaminants into leaves and release during litter fall, inability to assure clean-up below action levels in a short period of time, difficulty in establishing the vegetation due to toxicity at the site, and possible migration of contaminants off-site by macro pore flow or by binding with soluble plant exudates. Among other limitations, activities induced through phytoremediation are slower than the conventional approaches and effective at low concentrations.

These issues as mentioned above need to be addressed scientifically and require long-term monitoring of sites for it to be acceptable to the regulatory bodies.

Future Challenges

Though phytoremediation technologies are still primarily in research and development phase, some key technical hurdles that must be overcome are identification of more species that have remediative abilities, optimizing phytoremediation processes (appropriate plant selection and agronomic practices), better understanding of how plants uptake, translocate, and metabolize contaminants, identifying genes responsible for uptake/or degradation for transfer to appropriate high-biomass plants, decreasing the length of time needed for phytoremediation to work, devising appropriate methods for contaminated biomass disposal, particularly for heavy metals and radionuclides that do not degrade to harmless substances, and protecting wildlife from feeding of plants used for remediation.

R&D strategies require development of transgenic plants and molecular techniques approaches for genetic and physiological enrichment, removal and detoxification of toxic metals and metalloids like arsenic, chromium, lead, selenium, mercury, cadmium, etc. from contaminated soil, sediments and water. Phytochemical engineering for detoxification of contaminants is an emerging area in phytoremediation research. Development of engineering design concepts and their validation would be the challenge for experts working in this area.

REFERENCES

Arora, A. and Singh, P.K., 2004. Accumulation of Heavy Metals by *Azolla* Spp. MoEF Funded Project.

Juwarkar, A.S., Juwarkar Asha and Khanna, P., 2004. Use of Selected Waste Materials and Biofertilizers for Industrial Solid Waste Reclamation, Solid Waste: Assessment, Monitoring and Remediation, Twardowska, Allen, Kettrup and Lacy (Eds.), 911-48.

Juwarkar, Asha A., Singh, Sanjeev Kumar, Dubey Kirti and Nimje Milind, 2003. Reclamation of Iron Mine Spoil Dump Using Integrated Biotechnological Approach, National Seminar on Status of Environmental Management in Mining Industry, Banaras Hindu University, Varanasi, 17-18 January, 2003, 197-212.

Babu, C.R., 2002. Restoration of Degraded Ecosystems at Asola and Bhatti Wildlife Sanctuaries, DBT Funded Project.

MoEF Annual Report, 2001-02.

Schnoor, L.J., 2002. Phytoremediation of Soil and Groundwater, Technology Evaluation Report-TE-02-01, GWRTAC (Groundwater Remediation Technologies Analysis Centre) "E" Series of Reports. (http://www.gwrtac.org)

NEERI, 2000. Demonstration Project on Restoration of Manganese and Coal Mines Spoil and Fly Ash Dumps through Integrated Biotechnological Approach submitted to Department of Biotechnology, New Delhi.

TERI, 2000. Tata Energy Research Institute Report, Mycoorhizal Technology for Reclamation of Fly Ash Dumps submitted to NTPC, New Delhi.

Ram Prasad, 1999. *Technologies for Wasteland Reclamation*, Associated Publishing Co., New Delhi.

Rai, U.N. and Pal, A., 1999. Toxic Metals and Phytoremediation, Protect Environment Conserve Biodiversity Save Planet Earth.

Glass, D.J., 1998. (http://www.channell.com/dglassassoc/INFO/phytrept.htm)

Juwarkar, A.A., Juwarkar, A.S., Kulkarni, A.B. and Khanna, P., 1998. Integrated Biotechnological Approach for Sustainable Development of Wastelands, National Symposium on Role of Microbes in the Management of Environmental Pollution, Department of Microbiology, Punjab University, Chandigarh.

Juwarkar, A.S., Juwarkar, Asha, Pandey, V.S., Bal, A.S. and Malhotra, A.S., 1992. Restoration of Manganese Mine Spoil Dump Productivity using Pressmud, 2nd International Conference on Environmental Issues and Management of Waste in Energy and Mineral Production, Canada, 1-4 September.

Juwarkar, A., Juwarkar, A.S., Malhotra, A.S. and Jiwtode, S., 1991. Restoration of Microbial Activity in Mine Spoil Dumps Using Waste

Material and Specialized Cultures, International Conference of Mineral Development and Environment, Dhanbad.

Irshad Khan, 1987. *Reclamation of Salt Affected Soil*, Associated Publishing Co., New Delhi.

Forest Research Institute (FRI), Annual Report, 1977. Dehradun.

BIODIVERSITY

Biodiversity—Issues and Challenges

36

T.K. Ghosh

Preamble

Biodiversity is the variety of living organisms considered at all levels, viz. genetics through species, to higher taxonomic levels, and including the variety of habitats and ecosystems, as well as the processes occurring therein. Biodiversity is in fact more complex than species richness. There are in fact four levels of biodiversity: genetic diversity, species richness, ecosystem diversity and landscape diversity; and biodiversity also depend upon and encompass processes, as well as entities. These processes include, but are not limited to, biogeochemical cycles; biotic and abiotic disturbances; predator—prey, mutualistic, or parasitic relationships; migrations; competitive effects, and so forth. Thus, biodiversity includes all the entities of the living world at various levels of biological organization, plus the various things that those entities do (http://www.amazonia.org/Biodeversity/ALMA/A/ABDE_3.htm). India is one of the mega biodiversity centers in the world and has two of the world's 18 'biodiversity hotspots' located in the Western Ghats and in the Eastern Himalayas (Myers *et al.*, 2000). The forest cover in these areas is very dense and diverse and of pristine beauty, and incredible biodiversity.

One of the most serious threats to biodiversity, both locally and globally, is pollution. Pollution occurs when substances are released into an environment where they can have a harmful effect on living and non-living things. Human activities can pollute both on a local and global scale and may affect three of our most important natural resources, viz. water, soil and air.

Current Status

India has a rich and varied heritage of biodiversity, encompassing a wide spectrum of habitats from tropical rainforests to alpine vegetation and from temperate forests to coastal wetlands. India figured with two hotspots—the Western Ghats and the Eastern Himalayas—within 25 biodiversity hotspots (Myers *et al.*, 2000). The 2 hotspots that extend into India are The Western Ghats/Sri Lanka and the Indo-Burma region (covering the Eastern Himalayas); and they are included amongst the top eight most important hotspots. In addition, India has 26 recognized endemic centers that are home to nearly a third of all the flowering plants identified and described to date. According to an MoEF Report (1999), the country is estimated to have over 49,000 plant species and over 81,000 animal species representing 7% of the world's flora and 6.5% of its fauna.

Of the estimated 5–50 million species of the world's biota, only 1.7 million have been described to date (Groombridge and Jenkins, 2000), and the distribution is highly uneven. India contributes significantly to this latitudinal biodiversity trend. With merely 2.4% of the world's area, India accounts for 7.31% of the global faunal total comprising of 89,451 species (MoEF, 1999). Increasing human population, increasing consumption levels, and decreasing efficiency of use of our resources are some of the causes that have led to overexploitation and manipulation of ecosystems. Trade in wildlife, such as rhino horn, has led to the extinction of species.

India has an estimated 45,000 species of plants, 33% of which are native. About 3,000 to 4,000 of the plant species are believed to be threatened with extinction (http://www.india.punjabilit.com/land_and_resources.htm). India contains 172 species of animals, which are considered to be on the brink of extinction (threatened) by International Union for the Conservation of Nature and Natural Resources (IUCN). This corresponds to 2.9% of the world's total number of threatened species (Groombridge, 1993) comprising of 53, 69, 23 and 3 species of mammals, birds, reptiles and amphibians respectively. The red data book enlists the IUCN categories of plants and animals occurring in various parts of India (Nayar and Shastry, 1987).

To prevent such loss, the Government of India is setting up biosphere reserves in different parts of the country. The programme of Biosphere Reserve was initiated under the 'Man and Biosphere' (MAB) programme by UNESCO in 1971. These are multipurpose protected areas to preserve the genetic diversity in different ecosystems. Presently, there are 13 existing biosphere reserves in India, which altogether cover 54,809.42 sq. km area. Out of these 13 reserves, five are over 5,000 sq. km each (MoEF, 2000). Having witnessed the extinction of several hundreds of species in the 20th century, the Biological Diversity Act, 2002, recently enacted by the Indian Parliament, is a significant, indeed, a pioneering piece of legislation (Anonymous, 2002).

International Concerns

- As many as 50-200 species are lost from this planet every day, 20,000 to 70,000 per year and it takes between 2,000-100,000 generations for higher species to evolve. Nowhere does this trend present a clearer threat than in India, where human pressures increasingly jeopardize biodiversity.
- Lack of knowledge about species in the most biodiverse and at-risk areas.
- The fastest growing sectors of modern agriculture and modern medicine are those involving genetic engineering —without a sufficient gene pool (genetic diversity), there will be inadequate raw materials to work from (http://isis.csuhayward.edu/alss/geography/mlee/envt2000/biodivf99.htm).

Most Serious Concerns

- Anthropocentric point of view for preserving biodiversity—our limited knowledge of the existing biotic species (about 90% are still unidentified) requires urgent documentation and chemical evaluation, as some of these species might have economic, medical, aesthetic, recreational, scientific or ecological value at some future point.

- Earth-first point of view for preserving biodiversity—each species, as a unique product of millions of years of adaptation, has an inherent right not to be eliminated by human beings.
- The fastest growing sector in many developing countries is ecotourism, which relies upon the continued presence of pristine habitat.

Emerging Issues

- Long run bio resource availability and baseline assessment of available bio resource.
- Managing diverse biotic wealth with voluntary initiatives.
- Conflict and environmental security: communication and conflict resolution.
- Small scale biodiversity conservation and management of biodiversity hotspots.
- Policy aspects of bio resource management.
- National and Global information and finance dialogue.
- Public participation and planning for outcomes.

Inadequacy in Management and Control

Indian authorities have traditionally relied on policing to safeguard protected areas. Where funds have been available, they have used walls, gates, guards, and guns to protect parks and sanctuaries. However, these techniques have generally not proved effective in preventing exploitation of these areas by local communities, and they have often led to confrontation between wildlife authorities and villagers. As a result, there is a growing awareness in India, and in other countries as well, that another approach should be tried; one that secures willing cooperation by providing alternatives to reliance on the natural resources of protected areas.

Future Challenges

- Identification of ecologically fragile areas and sites in India, with clear provision for periodic updates.

- To determine interaction between biodiversity dynamics, ecosystem processes, and abiotic factors.
- Establishment of national repository centers for various life forms and use of information technology with networking and collaborative work.
- Identification of prioritized areas to initiate conservation programme.
- The economic valuation of natural resources and biodiversity is among the most pressing and challenging issues confronting today's environmental economists. More valuation studies need to be pursued to improve the information available to decision makers about the nature, magnitude and value on the impacts of human activities on the overall level of biodiversity.
- Application of mathematical models for evaluating biodiversity pattern. Based on modeling exercises, David Tilman from University of Minnesota believed that patchy distribution of resources leads to high diversity. However, it is yet to explain why biodiversity peaks at intermediate levels of productivity (http://www.findarticles.com/p/articles/m1511/isn9v15/ai15770746).

REFERENCES

Anonymous, 2002. Biological Diversity Act, as Passed by the Lok Sabha on 2nd December, 2002 and Rajya Sabha on 11th December, 2002, Parliament Secretariat, Government of India.

Groombridge, B. and Jenkins, M.D., 2000. Global Biodiversity: Earth's Living Resources in the 21st Century. UK: World Conservation Monitoring Center, 246.

Myers, N., Mittermeier, R.A., Mittermeier, C.G., da Fonseca, G.A.B. and Kents, J., 2000. Biodiversity Hotspots for Conservation Priorities, Nature, 403, 853-58.

MoEF, 2000. Annual Report 2000-01. New Delhi: Ministry of Environment and Forests, Government of India.

MoEF, 1999. National Policy and Macrolevel Action Strategy on Biodiversity, New Delhi, Ministry of Environment and Forests, Government of India, 74.

Groombridge, B. (Ed.), 1993. *The 1994 IUCN Red List of Threatened Animals*, IUCN, Gland, Switzerland and Cambridge, UK, lvi + 286.

Myers, N., 1988. Threatened Biotas: 'Hotspots' in Tropical Forests, *Environmentalist*, 8, 187-208.

Nayar, M.P. and Shastry, A.R.K. (Eds.), 1987. Red Data Book of Indian Plants, *Botanical Survey of India*, Calcutta, 1, 367.

Microbial Diversity

37

H.J. Purohit and Atya Kapley

Preamble

Microorganisms are the largest reservoir of genetic and biochemical diversity. They have been evolving for the last 3.8 billion years, and they inhabit virtually every environment, often thriving under extreme conditions of nutrient concentration, pH, salinity, pressure, and temperature (DeLong and Pace, 2001). Microbes drive the chemistry of life, do much of the biogeochemical cycling that keeps the world inhabitable, and also affect the global climate. Over billions of years, microbes have developed a wealth of functions that enable their survival in virtually every environmental niche, often where no other life forms exist. The microbial community is a black box and efforts are underway to understand the microbial wealth. The genomic information represented by such a "community genome" offers a tremendous resource for examining the extent and patterns of microbial genetic diversity and metabolic capabilities in the natural ecosystems.

Current Status

The reported culturable microbes comprise of only 1% of the total diversity (Pace, 1997; Torsvik *et al.*, 1998). Hence, very little is known about the spatial and temporal variability of microorganisms (Trevors, 1998; Ellis *et al.*, 2003). The limitations in studying microbial diversity do not arise only from methodological limitations, but also from a lack of taxonomic knowledge. In India, molecular techniques based on PCR have been used to overcome the limitations of culture-based methods (Kapley and Purohit, 2001; Moharikar *et al.*, 2003). These techniques have been used successfully to demonstrate the presence of bacteria in extreme environments, like hot springs and others (Kumar *et al.*, 2004; Higashi *et al.*, 2004; Kalin *et al.*, 1991; Hallberg and Johnson, 2003; Napolitano and Shain, 2004; White *et al.*, 2000). Lichens serve as useful indicators of environmental health. Similarly, several blue green algae help to sequester salt from water (Swaminathan, 2003). A tremendous flow of information has been created through various genome-sequencing projects worldwide. So far, 128 bacterial genome sequences have been completed and 391 are under way.

International Concerns

There are millions of species in the microorganismal domains viz. Bacteria and Archaea and only the basic principles governing their distribution and abundance in natural environments is reported. Despite the predominance of microbes during the course of life's history, general principles, theory of microbial evolution and ecology are not well developed. Hypotheses and models explaining the development are beginning to offer promising insights into the mechanisms governing prokaryote diversity and ecosystem function (Torsvik *et al.*, 2002). The development of molecular phylogenetics has recently enabled characterization of naturally occurring microbial biota without cultivation. Free from the biases of culture-based studies, molecular phylogenetic surveys have revealed a vast array of new microbial groups (Dojka *et al.*, 2000; Harris *et al.*, 2004). It is difficult to study the diversity of a group of microorganisms, when it is not understood how to categorize or identify the species present (Kirk *et al.*, 2004).

Most Serious Concerns

How are microbial species distributed and what controls their abundance? Are there geographic limits to distribution or are all species always present? Are the microbial species that have been well studied in the laboratory actually present, active, and important in nature; or have we been studying laboratory weeds? What is the best way to characterize microbial communities without identifying every species? These are a few questions that plague all microbiologists. The answer lies probably in the DNA and RNA, which is being tapped for such answers (Purohit *et al.*, 2005).

Emerging Issues

Knowledge about the metabolic and regulatory pathways of microbes and their communities will provide the foundation to begin understanding and using their remarkable capabilities, especially those related to environmental remediation, biogeochemical cycles, climate changes, and energy production. The use of molecular tools in microbial community analysis is fast replacing the conventional microbiological techniques, like biochemical tests and analysis based on growth media.

With the development of high throughput wet lab accessories and computational analysis, the curiosity for unexplored microbial wealth has got new dimensions. Instead of characterizing one single gene, the concept has grown to "How the gene pools survive under defined conditions?" With 95% or more unexplored microbial wealth around and their tremendous capacity to evolve at every time point, ETPs serve as rich genetic resources. The genetic capabilities of these ETPs find their role in generating new catabolic candidates with novel kinetics or affinities associated with them (Purohit *et al.*, 2005). The application of selected genomics tools for environmental impact analysis has been proposed so that a perturbation in ecosystem due to an industrial or any developmental activity could be analyzed at basic resource level (Purohit *et al.*, 2003).

Molecular Tools in Microbial Analysis

Randomly Amplified Polymorphic DNA (RAPD) Analysis: The analysis uses a single primer. The primer anneals to inverted repeating pattern in a target DNA at different positions; and under the defined amplification conditions it yields a set of amplicons, which are organism specific.

ARDRA (Amplified Ribosomal DNA Restriction Analysis): The amplified 16S rRNA gene can be differentiated amongst different genera using this tool. A single or a combination of restriction enzymes is used to digest the amplicon giving a genera digestion pattern.

DGGE (Denaturing Gradient Gel Electrophoresis): DGGE is an electrophoretic separation method based on differences in melting behavior of double stranded DNA fragments (Fisher and

Lermann, 1979). The principle of DGGE exploits that otherwise identical DNA molecule, which even if differs by only one nucleotide within a low melting domain will have different melting temperatures.

Terminal Restriction Fragment Length Polymorphism: In this analysis, PCR primers can be designed to be non-discriminating, such as amplifying nearly all 16S rDNAs using universal primers; and the 5' primer could be fluorescently labeled to tag the products. The PCR products could be digested with frequent cutters and analyzed (Dunbar *et al.*, 2000; Scala and Kerkhof, 2000).

Fluorescence *in situ* Hybridization Assay: One of the most powerful tools in modern microbial ecology is fluorescence *in situ* hybridization (FISH) (Christensen *et al.*, 1999; Bourne *et al.*, 2000). FISH allows the specific detection and enumeration of target populations directly in their natural environment without the need for cultivation.

Inadequacy in Management and Control

Substantial effort has been devoted to understand the distribution and role of microbial biomass in soils, sediments, and other important habitats in terrestrial and aquatic ecosystems. It has been observed that many of the microbial accumulation curves are linear or close to linear because of high diversity or small sample sizes or both; and hence the microbial diversity cannot be estimated (Hughes *et al.*, 2001). In such cases, the use of statistical and computational tools to understand the biological systems shall help to revolutionize the biological information domain.

Future Challenges

Sustainable development stands on a paradigm that minimum adverse environmental effects ensure long-term benefits. Hence, the conservation, enhancement, sustainable and equitable use of biodiversity should be accorded high priority in all national environment protection programmes. The information obtained from genomics and related technologies will have great impact on the future identification and development of new products like antimicrobial agents (Nes and Johnsborg, 2004). There is a need for more widespread use of such tools. Bioprospecting research design to identify novel metabolites must be rooted in the principle of equity in sharing benefits with the holders of traditional knowledge.

REFERENCES

Purohit, H.J., Kapley, A. and Raje, D.V., 2005. *Dynamics of Microbial Community: An Environmental Perspective, Microbial Diversity: Current Perspectives and Potential Applications*, T. Satyanarayana and B.N. Johari (Eds.), Kluwer Publishers, UK, 559-72.

Harris, J.K., Kelley, S.T. and Pace, N.R., 2004. New Perspective on Uncultured Bacterial Phylogenetic Division OP11. *Applied and Environmental Microbiology*, 70, 845-49.

Higashi, Y., Sunamura, M., Kitamura, K., Nakamura, K., Kurusu, Y., Jun-Ishibashi, I., Urabe, T. and Akihiko Maruyama, 2004. Microbial Diversity in Hydrothermal Surface to Subsurface Environments of Suiyo Seamount, Izu-Bonin Arc, Using a Catheter-type *in situ* Growth Chamber, FEMS Microbiology Ecology, 47, 327-36.

Kirk, J.L., Beaudette, L.A., Hart, M., Moutoglis, P., Klironomos, J.N., Lee, H. and Jack T. Trevors, 2004. Methods of Studying Soil Microbial Diversity, *Journal of Microbiological Methods*, 58, 169-88.

Kumar, B., Trivedi, P., Mishra, A.K., Pandey, A. and Palni, L.M.S., 2004. Microbial Diversity of Soil from Two Hot Springs in Uttaranchal Himalaya, *Microbiol. Res.* 159, 141-46.

Napolitano, M.J. and Shain, D.H., 2004. Four Kingdoms on Glacier Ice: Convergent Energetic Processes Boost Energy Levels as Temperatures Fall. Royal Society of London, Series B, *Biological Sciences*, 7, 271, Suppl 5, S273-76.

Nes, I.F. and Johnsborg, O., 2004. Exploration of Antimicrobial Potential in LAB by Genomics, *Current Opinion in Biotechnology*, 15, 100-04.

Ellis, R.J., Morgan, P., Weightman, A.J. and John C., 2003. Cultivation-Dependent and Independent Approaches for Determining Bacterial Diversity in Heavy-Metal-Contaminated Soil. *Applied & Environmental Microbiology*, 69, 3223-30.

Hallberg, K.B. and Johnson, D.B., 2003. Novel Acidophiles Isolated from Moderately Acidic Mine Drainage Waters, *Hydrometallurgy*, 71, 139-48.

Moharikar, A., Kapley, A., and Purohit, H.J., 2003. Detection of Dioxygenase Genes Present in Various Activated Sludge, *Environmental Science and Pollution Research*, 10, 373-76.

Purohit, H.J., Raje, D.V., Kapley, A., Padmanabhan, P. and Singh, R.N., 2003. Genomics Tools in Environmental Impact Assessment, *Environmental Science and Technology*, 37, 356A-63A.

Swaminathan, M.S., 2003. Bio-diversity: An Effective Safety Net Against Environmental Pollution, *Environmental Pollution*, 126, 287-91.

Torsvik, V., Ovreas, L. and Thingstad, T.F., 2002. Prokaryotic Diversity-magnitude, Dynamics and Controlling Factors, *Science*, 10, 296 (5570), 1064-66.

DeLong, E.F. and Pace, N.R., 2001. Environmental Diversity of Bacteria and Archaea, Systematic Biology, 50, 470-78.

Kapley, A. and Purohit, H.J., 2001. Tracking of Phenol Degrading Genotype, Environmental Science and Pollution Research, 8, 89-90.

Bourne, D.G., Holmes, A.J., Iversen, N. and Murrell, J.C., 2000. Fluorescent Oligonucleotide rDNA Probes for Specific Detection of Methane Oxidising Bacteria, FEMS *Microbiology Ecology,* 31, 29-38.

Dojka, M.A., Harris, J.K. and Pace, N.R., 2000. Expanding the Known Diversity and Environmental Distribution of an Uncultured Phylogenetic Division of Bacteria, *Applied Environmental Microbiology*, 66, 1617-21.

Dunbar, J., Ticknor, L.O. and Kuske, C.H., 2000. Assessment of Microbial Diversity in Four Southwestern United States Soils by 16S rRNA Gene Terminal Restriction Fragment Analysis. *Applied Environmental Microbiology*, 66: 2943-50.

Scala, D.J. and Kerkhof, L.J., 2000. Horizontal Heterogeneity of Denitrifying Bacterial Communities in Marine Sediments by Terminal Restriction Fragment Length Polymorphism Analysis. *Applied Environmental Microbiology*, 66, 1980-86.

White, D.M., Pilon, T.A. and Woolard, C., 2000. Biological Treatment of Cyanide Containing Wastewater, *Water Research*, 34, 2105-09.

Christensen, H., Hansen, M. and Sørensen, J., 1999. Counting and Size Classification of Active Soil Bacteria by Fluorescence *in situ* Hybridization with an rRNA Oligonucleotide Probe. *Applied Environmental Microbiology*, 65, 1753-61.

Torsvik, V., Daae, F.L., Sandaa, R.A. and Ovreas, L., 1998. Novel Techniques for Analysing Microbial Diversity in Natural and Perturbed Environments, *Journal of Biotechnology*, 64, 53-62.

Trevors, J.T., 1998. Bacterial Biodiversity in Soil with an Emphasis on Chemically-Contaminated Soils, *Water Air Soil Pollution*, 101, 45-67.

Pace, N.R., 1997. A Molecular View of Microbial Diversity and the Biosphere, *Science*, 276, 734-40.

Kalin, M., Cairns, J. and McCready, 1991. Ecological Engineering Methods for Acid Mine Drainage Treatment of Coal Wastes, Resources, Conservation and Recycling, 5, 265-75.

Fisher, S.G. and Lerman, L.S., 1979. Length-independent Separation of DNA Restriction Fragments in Two-dimensional Gel Electrophoresis Cell, 16, 191-200.